Java程序设计基础

（第2版）

■ 董　东　解建军　孙　慧　吴丽红　编著

清华大学出版社
北京

内 容 简 介

本书针对应用型本科计算机类专业教学目标，以"对象"的概念为核心，按照循序渐进的教学基本原则介绍 Java 面向对象程序设计基本思想、方法和技术，力图使学生较为轻松地理解"面向对象"程序设计范式，能够准确地理解 Java 程序，并能够应用 Java 语言解决实际问题。

全书共分 12 章。第 1 章介绍 Java 程序设计环境以及 Java 程序的一般结构，特别强调代码风格；第 2 章从一个简单的程序开始，介绍 Java 标识符、运算符、基本数据类型、简单的控制台输入输出、流程控制语句、数组；第 3 章介绍类的设计、对象的创建和访问；第 4 章介绍类与类之间的关系，包括继承和实现等；第 5 章介绍异常处理；第 6 章介绍常用的 API，如字符串处理、日期和时间处理等；第 7 章介绍 JCF 框架；第 8 章介绍泛型；第 9 章介绍反射；第 10 章是输入输出流，重点介绍磁盘文件的读写；第 11 章介绍线程和并发程序设计的基本技术；第 12 章介绍如何综合运用类、对象、异常、JCF 和输入输出流设计一个学生选课系统。

本书基于 Java SE 21。每章均提供电子版习题，并提供 PPT、源代码等配套资源。

本书可作为计算机类专业 Java 面向对象程序设计的入门教材，也可供专业技术人员参考。

版权所有，侵权必究。举报：010-62782989，beiqinquan@tup.tsinghua.edu.cn。

图书在版编目(CIP)数据

Java 程序设计基础 / 董东等编著. -- 2 版. -- 北京：清华大学出版社，2025.1.
（清华开发者学堂）.--ISBN 978-7-302-68042-0

Ⅰ. TP312.8

中国国家版本馆 CIP 数据核字第 2025VP2156 号

责任编辑：张　玥　薛　阳
封面设计：吴　刚
责任校对：徐俊伟
责任印制：宋　林

出版发行：清华大学出版社
网　　址：https://www.tup.com.cn，https://www.wqxuetang.com
地　　址：北京清华大学学研大厦 A 座　　　　邮　编：100084
社 总 机：010-83470000　　　　　　　　　　　邮　购：010-62786544
投稿与读者服务：010-62776969，c-service@tup.tsinghua.edu.cn
质量反馈：010-62772015，zhiliang@tup.tsinghua.edu.cn
课件下载：https://www.tup.com.cn,010-83470236

印 装 者：三河市龙大印装有限公司
经　　销：全国新华书店
开　　本：185mm×260mm　　　　印　张：20.5　　　　字　数：512 千字
版　　次：2017 年 12 月第 1 版　2025 年 1 月第 2 版　　印　次：2025 年 1 月第 1 次印刷
定　　价：69.50 元

产品编号：106669-01

前言

  面向对象程序设计已经成为当前桌面、服务器、移动应用开发的主流技术。以面向对象思想为核心的 Java 语言成为最受欢迎的语言之一，Java 程序已经广泛运行在各类平台和设备上。由于应用"面向对象"的思维模式解决问题是对人们在工作、生活中解决问题的一个自然的抽象，所以通过面向对象程序设计让计算机实现问题求解也就变得较为容易，尤其在复杂的应用场景中，面向对象程序设计降低了开发的难度，提高了程序的可理解性和可维护性。

  通过本书的学习，读者可以逐渐领会面向对象程序设计的基本思维模式，能够运用 Java 语言编写面向对象的程序解决具体问题。本书试图满足如下目标。

  (1) 强调面向对象思维方式的重要性。引导学生对现实生活中的对象及其协作来解决问题的场景进行抽象，理解类与对象间的关系，理解如何通过对象间的消息实现问题求解。

  (2) 强调提高问题求解能力。语言仅仅是表达思想的工具，学习语言的目的是有效地表达如何实现问题求解。引导学生学会抽象、封装，引导学生理解问题求解的时间和空间需求。

  (3) 强调程序设计风格。程序源代码不仅仅被计算机阅读，而且也被人阅读。当完成一部分源代码后，评审人员会阅读；当程序发布后发现了缺陷，代码维护人员也会阅读代码定位问题。引导学生建立代码风格意识，写出"赏心悦目"的源代码。

  在过去的 20 多年里，笔者一直从事 Java 语言的应用开发、教学和研究工作，其中包含 5 年的 Java 语言程序设计的双语教学。也从事后续课程"编译原理"的教学。通过本书，试图反映在教学中遇到的共性问题，比如空指针异常、文件路径、把语句写在类体中而不是方法体中以及在研究中发现的问题，比如 Java SE API 的使用频率和用法模式，以增强本书的实用性。

  Java SE (Java Platform，Standard Edition) 用以开发和部署桌面或服务器应用程序。从 Java SE 5 到 Java SE 21，出现了很多新的语言特征，如 switch 语句的标签规则、文本块、局部变量类型推理、try-with-resources

语句、Lambda 表达式等。本书试图反映这些新特征。书中所有源代码均在 Oracle OpenJDK 21 下编译通过并使用 Checkstyle 完成程序设计风格审计。

通过清华大学出版社网站可免费获取教学大纲、PPT 和样例源码。本书提供了在线题库和在线测验系统，读者扫描封底的作业二维码后可激活章节测验权限，扫描书中的二维码即可实现在线做题。教师可登录网站（app.qingline.net）领取本书配套的习题和作业系统，系统提供了组班、作业布置、在线编程、自动批阅、学情分析导出等功能。

感谢教育部"编译课程虚拟教研室"和河北师范大学对我教学研究的支持，以及我所教授的计算机科学与技术专业本科学生提出的有价值的反馈和建议。学生在课堂上和上机实验过程中提出的问题都对本书写作与习题设计产生了影响。感谢 Java 程序设计网络资源的所有贡献者。从这些贡献者的图文以及视频资料笔者也受益很多。

河北师范大学计算机与网络空间安全学院解建军、孙慧和吴丽红是"Java 面向对象程序设计"课程组成员，分别参与了部分章节的修订。董东负责统稿和全面修订。

董 东

2024 年 6 月于河北师范大学

目录

第 1 章　Java 简介　/1

1.1　面向对象的程序设计　/2
1.2　开发环境　/5
1.3　在命令提示符窗口设计程序　/7
1.4　使用 jGRASP 设计程序　/10
1.5　使用 Eclipse 设计程序　/14
1.6　Java 程序结构　/20
1.7　代码风格　/21
　　1.7.1　命名约定　/21
　　1.7.2　留白　/21
　　1.7.3　块风格　/22
1.8　注释　/22

第 2 章　Java 语言基础　/26

2.1　标识符和保留字　/26
2.2　基本数据类型　/27
2.3　字面量　/28
2.4　变量　/30
2.5　运算符　/32
　　2.5.1　赋值运算符　/32
　　2.5.2　算术运算符　/33
　　2.5.3　关系运算符　/34
　　2.5.4　逻辑运算符　/35
　　2.5.5　条件运算符　/36
　　2.5.6　位运算符　/36
　　2.5.7　运算符的优先级　/38

2.6 表达式和语句　/39
2.7 控制台输入和输出　/39
2.8 控制结构　/41
 2.8.1 顺序结构　/41
 2.8.2 分支结构　/41
 2.8.3 循环结构　/49
 2.8.4 分支语句　/52
2.9 数组　/53
 2.9.1 数组的概念　/53
 2.9.2 数组的声明　/53
 2.9.3 数组的访问　/54
 2.9.4 二维数组　/56

第3章 类和对象　/59

3.1 类的声明　/59
3.2 创建对象　/62
3.3 访问对象　/63
3.4 对象的字符串表示　/65
3.5 方法的调用和返回　/68
3.6 重载　/73
3.7 类变量和实例变量　/74
3.8 静态方法和实例方法　/78
3.9 命令行参数　/79

第4章 继承　/81

4.1 通过继承共享祖先的特征　/81
4.2 父类和子类的构造方法　/84
4.3 覆盖实例方法和隐藏静态方法　/87
4.4 上转型和下转型　/90
4.5 抽象类和抽象方法　/92
4.6 保留字 final　/94
4.7 接口　/95
4.8 多态　/100
4.9 源代码的组织和访问控制　/105
4.10 Object 类　/111
 4.10.1 toString 方法　/111
 4.10.2 equals 方法　/112

4.10.3　hashCode 方法　/115
4.11　枚举　/116

第 5 章　异常　/119

5.1　异常的抛出与捕获　/119
5.2　处理异常　/123
5.3　自定义异常　/127

第 6 章　实用类　/132

6.1　字符串　/132
6.2　正规表达式　/141
6.3　编辑字符串　/146
6.4　字符对象　/148
6.5　数值对象　/149
6.6　数学运算 API　/150
　　6.6.1　Math 类　/150
　　6.6.2　Random 类　/151
　　6.6.3　BigInteger 类　/153
6.7　日期和时间　/154
　　6.7.1　LocalDate 类　/154
　　6.7.2　LocalTime 类与 LocalDateTime 类　/156
　　6.7.3　时间戳　/159
　　6.7.4　Date 类　/160
　　6.7.5　Calendar 类　/161
6.8　Arrays 类　/162

第 7 章　JCF 框架　/164

7.1　JCF 框架简介　/164
7.2　Set 接口和实现类　/168
7.3　List 接口　/174
　　7.3.1　ArrayList 实现类　/175
　　7.3.2　LinkedList 实现类　/178
　　7.3.3　Collections 类　/179
7.4　队列接口和实现类　/186
7.5　栈　/187
7.6　Map 接口和实现类　/189

7.7 流 /195

第 8 章 泛型 /203

8.1 概述 /203
8.2 泛型类 /205
8.3 泛型接口 /206
8.4 泛型方法 /209

第 9 章 反射 /212

9.1 Class 类 /212
9.2 实例化对象 /214
9.3 查询类的成员 /215
9.4 调用成员方法 /218

第 10 章 输入输出流 /219

10.1 文件与文件夹管理 /219
10.2 流 /225
    10.2.1 字节流 /226
    10.2.2 缓冲字节流 /229
    10.2.3 数据流 /232
    10.2.4 字符流 /234
10.3 Scanner 类和 PrintWriter 类 /237
10.4 对象串行化 /243
10.5 字符集和 Unicode /245
10.6 记录 /246

第 11 章 多线程程序设计 /250

11.1 进程和线程 /250
11.2 创建线程 /253
11.3 线程状态 /256
11.4 线程池 /261
11.5 线程安全的程序设计 /265
    11.5.1 与时间有关的错误 /265
    11.5.2 volatile 保留字 /268
    11.5.3 synchronized 保留字 /270

11.5.4 计数器 Adder　　/273
11.6 获取子线程的返回结果　　/275
11.7 BlockingQueue　　/276

第 12 章　学生选课系统　/282

12.1 需求分析　　/282
12.2 架构设计　　/283
12.3 详细设计　　/284

附录 A　Unicode Basic Latin 字符　/297

附录 B　jGRASP 调试和展演　/301

附录 C　jGRASP 单元测试　/310

参考文献　/315

第 1 章 Java 简介

进入 20 世纪 90 年代后,计算机的发展开始呈现出跨平台、网络化、嵌入式的特征。计算机硬件及网络环境的变化对计算机语言提出了新的要求。例如,能够运行在各种消费类电子产品上,如手机、平板电脑、洗衣机、空调、微波炉等;能够支持多线程;能够以"面向对象"的范式(paradigm)编写程序,以简化大规模复杂应用的开发等。市场需求使得 Sun 公司研发并于 1995 年发布了 Java 语言。1996 年 1 月,Sun 公司发布了 Java 的第一个开发工具箱 JDK 1.0 版本。Java 的设计理念强调"一次编写,到处运行"(Write Once,Run Anywhere),它通过 Java 虚拟机(Java Virtual Machine,JVM)和字节码(Bytecode)机制实现了跨平台兼容性。

1999 年 6 月,Sun 公司继续发布了第二代 Java 平台(简称 Java 2)的 3 个版本,分别是:J2SE(Java 2 Standard Edition),用于桌面系统应用开发;J2EE(Java 2 Enterprise Edition),用于企业应用开发;J2ME(Java 2 Micro Edition),用于移动终端应用开发。2004 年 9 月,J2SE 1.5 发布。该版本是从 1996 年的 1.0 版本以来最重大的更新,包括泛型支持、自动装箱、增强 for 循环、枚举类型等。为了强调该版本的重要性,J2SE 1.5 更名为 J2SE 5.0。2005 年 6 月,Sun 公司发布 Java SE 6。从此,各种 Java 版本中不再有"2":J2SE 改为 Java SE;J2EE 改为 Java EE;J2ME 改为 Java ME。2009 年,Sun 公司被甲骨文(Oracle)公司收购。2014 年,甲骨文公司发布了 Java 8。自 Java SE 8 以来,Java 持续进行重大更新,添加了许多新特性,如 Lambda 表达式、标签规则、局部变量类型推理、记录等。Java 已经广泛应用于解决各种领域的应用问题。

Java SE 是 Java EE 和 Java ME 的基础,三者的关系如图 1-1 所示。Java EE 已经被转移至 Eclipse 基金会,并更名为 Jakarta EE。

图 1-1　Java SE、Java EE 和 Java ME 之间的关系

OpenJDK 是 Java 标准实现的主要开源参考实现,其他活跃的 OpenJDK 分支和发行版还有 AdoptOpenJDK、Amazon Corretto、Azul Zulu、Red Hat OpenJDK 等。

1.1 面向对象的程序设计

Java 是一种面向对象的高级程序设计语言。"面向对象"是一种程序设计范式:在运行时刻通过对象间的合作实现问题求解。某对象 p 上预定义了若干功能,称为"方法";另外一个对象 q 访问对象 p 的功能称为"方法调用"或者"消息传递"。具有相同特征的对象的抽象称为"类",Java 面向对象程序设计的一般过程就是先设计类,然后由类来创建对象。所以从源代码的角度看,Java 程序是一组类的集合。一般认为:

面向对象 = 类 + 对象 + 消息 + 继承

类的成员有成员变量、成员方法和构造方法等。成员变量用来存储对象的属性;成员方法用来实现对象的功能;而构造方法则用来创建对象。保留字 class 用来定义类。通常一个类保存为一个单独的源文件,扩展名为 java,如图 1-2 所示。一辆小汽车的属性有发动机排量(displacement)、型号(model)、车身颜色(color)、行驶速度(speed)等,汽车功能有加速(accelerate)、鸣笛(honk)、刹车(brake)等。把属性定义为类 Car 的成员变量,把功能定义为成员方法,再增加构造方法就得到小

图 1-2 小汽车

汽车的类定义 Car。类 Car 的定义保存在文本文件 Car.java 中,如源文件 1-1 所示。注意源文件 Car.java 的基本名 Car 与类名 Car 必须严格相同(大小写也要相同),因为 Java 对大小写敏感。

成员变量前的保留字 private 表示该变量是"类私有"变量,仅允许类内的其他成员访问。double 和 int 是表示双精度浮点数和 4 字节整数的基本数据类型。String 是预定义的字符串类,用来创建字符串对象。构造方法的名字 Car 与类的名字严格相同。

Java 语言的程序约定从 main 方法开始执行。源文件 1-2 的类 MyCar 中有 main 方法。该方法使用类 Car 声明了变量 myCar 准备引用 Car 类型的对象(第 3 行);使用保留字 new 调用构造方法创建了 Car 类的对象并将该对象的引用赋值给变量 myCar(第 4 行);最后通过对象的引用 myCar 调用对象上的方法 honk,即向 myCar 所引用的对象发送消息 horn,让其执行 honk 方法。程序的输出如下:

嘀嘀

源文件 1-1 Car.java

```
1 class Car {
2
3     //成员变量
4     private double displ; //displacement
5     private String model;
```

```
6      private String color;
7      private int speed;
8
9      //构造方法
10     Car(double displ, String model, String color) {
11         this.displ = displ;
12         this.model = model;
13         this.color = color;
14     }
15
16     //成员方法
17     int accelerate(int increment) {
18         this.speed = this.speed + increment;
19         return this.speed;
20     }
21     void honk() {
22         System.out.println("嘀嘀");
23     }
24
25     //...
26 }
```

源文件 1-2　MyCar.java

```
1 public class MyCar {
2     public static void main(String[] args) {
3         Car myCar;
4         myCar = new Car(1.6, "307", "银灰");
5         myCar.honk();
6     }
7 }
```

创建对象时发生了三件事情：实例化对象、初始化对象和获取对象引用。实例化对象就是根据类中成员变量的定义为对象分配存储以存放成员变量的值，这些值形成了对象的状态；初始化对象就是执行构造方法设置成员变量的值；获取对象引用就是把对象的引用赋值给已经声明的引用类型变量。源文件 1-2 的第 3～5 行执行后的对象在内存中的布局如图 1-3 所示。

由于在类 Car 中定义了 4 个成员变量，所以实例化的对象由 4 个成员变量组成。初始化后这 4 个成员变量的值分别是 1.6、字符串"307"的引用、字符串"银灰"的引用和 0。对象和值放置在矩形框中。字符串也是对象，所以在 model 和 color 变量中存储的是对象引用，图 1-3 中，从表示存储空间的正方形出发，末端有箭头的实线表示引用。变量 myCar 中存放的是所创建的对象的引用。对象里没有存储方法，因为从一个类可以创建多个对象，这些对象的变量取值可以不同但可以执行的方法是相同的，所以共享类中的方法。

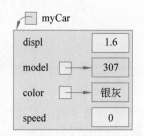

图 1-3　变量 myCar 及其引用的对象（截取自 jGrasp 展板）

类就是依据其创建的所有对象的共同特征的抽象。使用 UML 类图可视化这个抽象的类。类图是统一建模语言（Unified Modelling Language，UML）中重要的图之一。它描述类的成员以及类之间的关系。在类图中，矩形框表示"类"。一个矩形框分成了三个隔间，自上而下分别描述类的名字、类的属性以及类的操作。类的操作定义了类的行为特征，描述能够完成的功能。除了名字隔间必须显示，其他两个隔间可以不显示。图 1-4 的类图描述了类模型 Car。类 Car 的属性有发动机排量（displ）、型号（model）、车身颜色（color）、行驶速度（speed）等；行为有加速（accelerate）、鸣笛（honk）等。属性名字前的-表示该属性是"类私有"属性；行为前的~表示该行为是"包私有"行为。

类之间的关系有：继承（inheritance）、关联（association）、实现（realization）和依赖（dependency）。

继承关系通常用来对现实世界中的"is-a"关系进行建模。比如"汽车是交通工具"，即"A car is a vehicle"。"is-a"关系称为继承关系。如果类 A 被类 B 继承，即 B is a A，那么 A 称为 B 的父类，B 称为 A 的子类。"A car is a vehicle"的意思是类 Car 继承了类 Vehicle（Java 规范中类名首字母大写）。

继承就使得在定义和实现一个类的时候，可以在一个已经存在的类的基础之上来进行，并可以加入若干新的属性和行为，或修改原来的属性和行为使之更适合子类需要。继承是一种代码共享机制。提高了软件的可重用性和可扩展性。

在 UML 类图中，使用末端是空心三角的实线连接具有继承关系的两个类，末端指向父类，如图 1-5 所示。

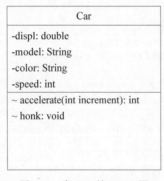

图 1-4　类 Car 的 UML 图

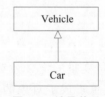

图 1-5　继承关系

实现关系意思是一个类实现了另外一个类中声明的行为。在 UML 类图中使用末端是空心三角箭头的虚线表示。接口（interface）和接口的实现类的配合体现了开闭原则：对扩展开放，对修改封闭。也就是说，要在不改动当前类的情况下增加新的功能。在接口中展示了简单的 API 接口；在实现类中提供复杂的功能实现。例如在驾驶员与汽车的关系中，驾驶员看到的是方向盘、刹车踏板、油门踏板、车速仪表盘等；而汽车如何传导信号给刹车片刹车、发动机如何加速等实现细节被隐藏了。

车主和小汽车之间的产权关系称为关联，这种关联关系在 UML 类图中使用连接两个类之间的实线表示；依赖关系意思是一个对象的改变会强迫另外一个对象进行改变，尽管这两个对象之间没有显式的关联。依赖关系使用末端是开箭头的虚线表示。

封装性是面向对象的程序设计的主要特征之一。对访问者隐藏对象的状态或内部表示称为"封装"。在源文件 1-1 的第 4～7 行使用关键字 private 限制成员变量仅能在本类中被访问就实现了封装。private 称为"访问修饰符",在 UML 类图中使用"-"表示。如果在成员前面没有访问修饰符,称为"包私有",在 UML 类图中使用"～"表示。封装的目标就是要降低类与类之间的耦合度,减少由于类与类之间的相互依赖性而带来的负面影响。面向对象的封装性把对象封装成一个高度自治和相对封闭的个体,对象状态(属性)由这个对象自己的行为(方法)来读取和改变。

通过相同类型的引用把相同的消息发送给不同的对象产生不同的效果,这种现象称为"多态"。多态是在运行时刻才确定对象的行为的机制。这样,不修改程序代码就可以改变程序运行时所绑定的具体代码,让程序可以选择多个运行状态。多态性增强了程序的可扩展性。

通常在遇到具体问题时,首先应明确问题,然后从对象及其之间的消息传递角度分析问题,抽象出创建对象所需要的类,这些类之间可能存在以上说明的关系,类及其关系形成类模型。定义了类模型之后再使用类创建对象,通过对象间的协作实现问题求解。这是面向对象程序设计解决具体问题的一般思路。如图 1-6 所示。面向对象的程序设计过程是一个迭代的过程:类模型往往经过多次迭代演化。

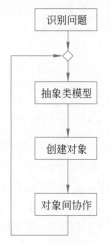

图 1-6 面向对象程序设计的过程

1.2 开发环境

Java 语言的开发环境包括 Java 开发工具箱(Java Development Kit,JDK)和集成开发环境(Integrated Development Environment,IDE)。JDK 包含了基本的程序设计工具,如编译器(compiler)、解释器(interpreter)、装入器(loader)、调试器(debugger)等,以及设计和运行程序所需的基本类库(如完成输入输出的 API、常用数学运算的方法实现等)。IDE 则把各种工具集成在一起,甚至包括代码补全、注释生成等功能的 AI 助手,提高了开发效率,如 Eclipse、Visual Studio Code、IntelliJ IDEA 等。

设计 Java 程序的过程一般如下。

(1) 设计若干类,并将其分别保存为相应的.java 文本文件,即扩展名为.java,文件名与类名相同。

(2) 编译源文件,生成相应的以.class 为扩展名的文件。这些.class 文件称为字节码文件。

(3) 启动 Java 虚拟机(JVM),将字节码文件装入虚拟机运行。

(4) 如果在编译时刻或者运行时刻发生错误,则修改源文件,重新编译和运行。

当程序经过测试和验收后,就可以进行具体应用部署了。

仅用于运行 Java 程序的计算机中无须安装 JDK,只需安装 Java 运行环境(Java Runtime Environment,JRE)。JRE 提供了运行 Java 程序所需的类库,解释并运行.class 文件的 Java 虚拟机以及其他工具。JRE 以及 JDK 统称为 Java 平台,如图 1-7 所示。

图 1-7　Java 平台

　　从官网下载 JDK 之前首先查看准备安装 JDK 的计算机系统的软件操作系统平台,以选择相应的安装软件版本。以 JDK 17 为例,在浏览器的地址栏输入网址:https://www.oracle.com/cn/java/technologies/downloads/♯jdk17-windows,单击下载链接下载 jdk-17_windows-x64_bin.exe。下载完成后在 Windows 资源管理器中双击,按照安装向导的提示安装。安装 JDK 的目标文件夹称为 Java home,可设置为 C:\Program Files\Java\jdk-17。

　　安装完成后,在 Java home 文件夹中会有 bin、conf、include、jmods、legal、lib 以及其他文件,如图 1-8 所示。前文提及的用于编译的工具 javac.exe,用于装入运行.class 文件的工具 java.exe 等都在 bin 文件夹中。

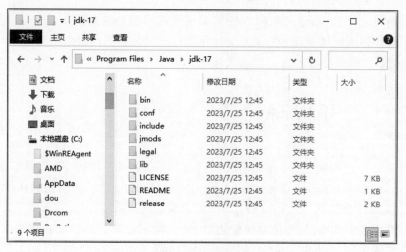

图 1-8　Java home

　　由于编译命令 javac 和装入运行命令 java 都是 Windows 操作系统的外部命令,为了能在 Windows 的任何文件夹中执行这两个命令,需要通过设置环境变量 PATH 告诉 Windows 操作系统到哪个文件夹中装入和执行命令。设置 Windows 10 环境变量 PATH 的步骤如下。

　　(1) 在资源管理器中打开 Java home 文件夹中的 bin 文件夹。

　　(2) 设置资源管理器显示地址栏以及在地址栏中显示完整路径,并从地址栏中复制 bin 文件夹的完整路径,如 C:\Program Files\Java\jdk-17\bin。

　　(3) 在桌面上右击"我的电脑",在弹出的快捷菜单中选择"属性",在弹出的"系统属性"对话框中选择"高级"选项卡,单击"环境变量"按钮。

　　(4) 在弹出的"环境变量"对话框的"系统变量"列表框中选择变量 PATH,单击"编辑"按钮。

（5）在"编辑系统变量"对话框中单击"变量值"文本框使其获得输入焦点，在键盘上按 Home 键使插入定位在行首。

（6）使用 Ctrl+V 组合键粘贴 bin 文件夹的完整路径 C:\Program Files\Java\jdk-17\bin，再键入英文标点符号中的分号"；"与其他路径隔开。

（7）单击"确定"按钮关闭所有对话框。

设置 Windows 11 环境变量 PATH 的步骤如下。

（1）在资源管理器中打开 Java home 文件夹。

（2）设置资源管理器显示地址栏以及在地址栏中显示完整路径，并从地址栏中复制 bin 文件夹的完整路径，如 C:\Program Files\Java\jdk-17。

（3）打开"控制面板"，在其中单击"系统和安全"，在弹出的对话框中再单击"系统"，继续单击"高级系统设置"，出现"系统属性"对话框，在"高级"选项卡中单击"环境变量"按钮。

（4）在弹出的"环境变量"对话框的"系统变量"列表框中单击"新建"按钮，在弹出的"新建系统变量"对话框中输入环境变量名 JAVA_HOME，环境变量值 C:\Program Files\Java\jdk-17，如图 1-9 所示，单击"确定"按钮。

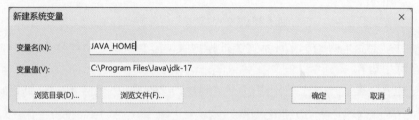

图 1-9　新建系统变量

（5）继续在"环境变量"对话框的"系统变量"列表框中选择变量 Path，单击"编辑"按钮，在"编辑系统变量"对话框中单击"新建"按钮，在光标插入点输入%JAVA_HOME%\bin，单击"确定"按钮关闭所有对话框。

当 Windows 操作系统从当前文件夹中找不到所要运行的外部命令时，就会依次从 PATH 环境变量以分号隔开的文件夹中去寻找相应的命令。定义环境变量 JAVA_HOME 的目的在于方便在多个环境变量值中引用。%JAVA_HOME%就是 C:\Program Files\Java\jdk-17。

最后在命令提示窗口测试 JDK17 是否安装成功。按下 Windows 徽标键的同时按下 R 键，打开 Run 窗口，输入命令 cmd，打开命令提示窗口。在命令提示窗口中输入 java -version，按 Enter 键，如果能看到命令输出版本信息，即表明安装成功。接下来，就可以使用 JDK 进行 Java 程序设计了。

1.3　在命令提示符窗口设计程序

使用命令进行 Java 程序设计包括编辑、编译、运行程序三个主要步骤。编译把源文件变换为字节码文件，扩展名为.class。Java 通过启动虚拟机运行字节码文件。编译时检查词法和语法错误，比如括号不匹配，缺少分号等。根据编译器输出的错误提示，在源程序中改正这些错误后重新编译,反复修改、编译,直到编译成功。如果运行时程序异常终止或者输出错误结果,则再修改源程序、重新编译和运行。对于运行时刻的错误,往往需要调试才能发现错误原因和位置。

javac 命令负责编译 Java 源代码,转换成字节码。java 命令负责启动 Java 虚拟机(JVM),装入和解释执行字节码。Java 字节码文件是跨平台的,但 javac 命令以及 JVM 不可移植,是针对特定平台的。所以首先使用文本编辑器,如 Notepad3(https://www.rizonesoft.com/),编辑若干类,并将其保存为与类的名字相同的.java 源文件;然后使用命令 javac 编译这些.java 源文件,生成相应的.class 字节码文件。如果在编译时刻或者运行时刻发生错误,则返回源文件进行修改,重新编译和运行。这个过程如图 1-10 所示。

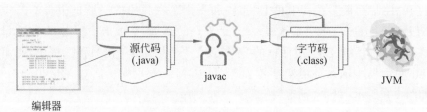

图 1-10　Java 程序设计的步骤

下面通过一个简单的例子来展示这个过程。假设需要设计一个计算器完成整数的加法和减法两个算术运算。

首先打开命令提示符方式;创建用于存放程序的文件夹,比如 work;然后启动文本编辑器 Notepad3,输入源文件 1-3 中所示的源代码。

源文件 1-3　Calculator.java

```
public class Calculator{
    private int a;
    private int b;

    void setA(int x){
        a = x;
    }
    void setB(int y){
        b = y;
    }
    int add(){
        return a + b;
    }
    int minus(){
        return a - b;
    }
}
```

把源代码保存为 work 文件夹中的 Calculator.java。然后在命令提示符窗口输入命令:javac Calculator.java,如图 1-11 所示。

使用 dir 命令查看生成的 Calculator.class 文件,如图 1-12 所示。

此时还不能启动.class 文件运行,因为 Calculator 类仅仅声明了两个用于运算的整型变量以及这两个变量上的操作 add 和 minus。谁来"调用"add 或者 minus 呢?要想调用这两

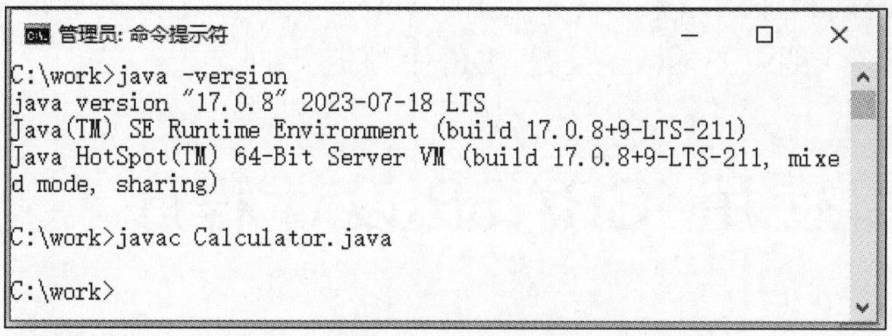

图 1-11　编译 Calculator.java

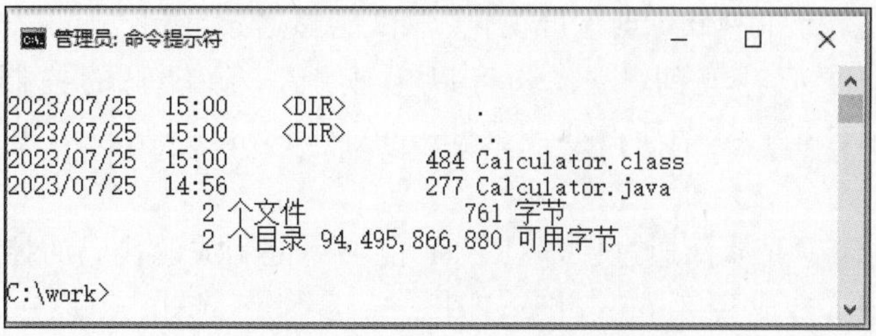

图 1-12　查看 work 文件夹中的字节码文件

个操作，就得先以 Calculator 为模板创建对象，然后让对象执行这些操作。所以还要写一个类，用以创建对象和向对象发送消息。再次使用 Notepad3 新建 Java 类，录入如源文件 1-4 中的源代码。

源文件 1-4　Test.java

```
public class Test{
    public static void main(String[] args){

        Calculator c = new Calculator();

        c.setA(10);
        c.setB(2);

        int result = c.add();

        System.out.println(result);
    }
}
```

保存源程序为 Test.java。使用如下命令编译：

javac Test.java

最后，使用 java 命令运行程序：

java Test

在这个例子中，Java 程序由两个类的定义组成：Calculator 和 Test。事实上，设计 Java 程序，就是设计一组完成任务的类。类由关键字 class 进行定义。后面花括号{}括起来的部分是类体。类体中声明的变量和方法都是类的成员。Java 程序约定从 main 方法开始执行。本例中的 main 方法使用保留字 new 创建了 Calculator 类的实例 c，通过让 c 执行 setA 方法和 setB 方法分别设置了成员变量 a 和 b 的值。然后让对象 c 执行 add 方法完成成员变量 a 和 b 相加,并返回相加的结果。最后使用 System 类中预定义的标准输出对象 out 上的 println 方法输出运算结果。System 类是 JDK 中预定义的类。

1.4 使用 jGRASP 设计程序

jGRASP 是奥本大学(Auburn University)使用 Java 开发的轻量级集成开发环境。其最大特点是能够和 JDK 捆绑安装,并集成了单元测试、源代码风格检查等工具,一键安装并自动设置环境变量。另外一个特点是可视化,不仅能够通过"展板(Canvas)"在调试模式中动态地、可视化地展示对象的状态,还可以生成 Java 源代码的控制结构图(Control Structure Diagram,CSD)、复杂性概要图(Complexity Profile Graph,CPG)以及 Java 项目的 UML 类图。

当键入完成源代码并编译通过后,jGRASP 能够一键生成源代码的控制结构图,使得源代码可以被叠起(fold)或铺平(unfold)。图 1-13 展示了单击工具条"生成控制结构图(Generate CSD)"按钮后类 Car 的控制结构图。每个方法左侧的竖线与类的左侧竖线在视觉上展示了包含关系。双击竖线左侧空白区域可叠起该竖线右侧的语法单元;双击"加号"可铺平语法单元。

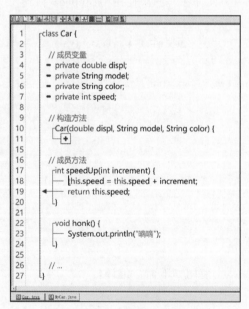

图 1-13 控制结构图

运行 jGRASP 设计 Java 程序需要 Java 11 或更高版本的 JDK。通常安装与 OpenJDK 捆绑的 jGRASP 发行版。

在 Windows 浏览器中打开 https://www.jgrasp.org/index.html,单击超级链接 download,打开下载页面,在"jGRASP 2.0.6_15(January 25,2024) - requires Java 11 or higher Bundled with Java (OpenJDK) 21.0.2, Checkstyle 10.9.3, and JUnit 4.13.2"区域单击 jGRASP Bundled X64 exe 按钮,下载 jGRASP 的 Windows 64 位捆绑版本。双击该 exe

文件启动安装过程。

安装完成后在 Windows 桌面出现 jGRASP 快捷方式，双击该快捷方式运行 jGRASP。默认以单进程模式运行。如果又一次双击快捷方式，jGRASP 会弹出当前正在运行的进程。Windows 系统托盘中提供了一个名为 jGRASP Control Shell 的 Windows 应用程序。这将显示 jGRASP 输出（如发生崩溃后的堆栈转储），并允许手动终止 jGRASP。

jGRASP 的用户界面如图 1-14 所示。首次运行的第一件事就是把源代码编辑器的字体设为"微软雅黑"或其他中文字体。单击 Settings 菜单中的 Font 菜单项，弹出"设置"对话框，如图 1-15 所示。

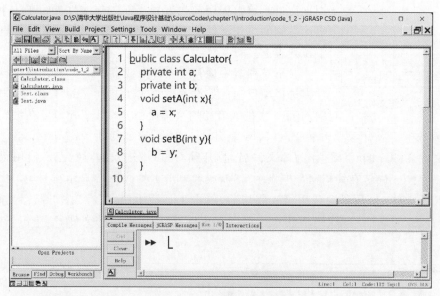

图 1-14　jGRASP 用户界面

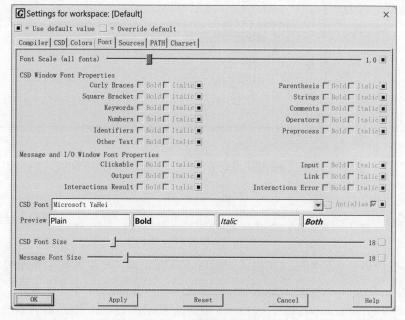

图 1-15　"设置"对话框

在 CSD Font 下拉列表中选择 Microsoft YaHei，单击 OK 按钮。每个项目尾部方框内的黑点表示使用默认值。如果不想使用默认值，那么首先单击黑点，取消默认值，再选择另外的值。例如设置编辑器的字号为 18，那么单击 CSD Font Size 末端的黑点，然后在滑块右侧单击，字号值就改为了 18，单击 OK 按钮。修改字体的目的是让 jGRASP 正确显示汉字字符。

CSD 窗口就是源代码的编辑窗口。由于能够在该窗口中显示控制结构图，所以也称为 CSD 窗口。

在 jGRASP 用户界面左侧的 Browse 标签页中是磁盘文件浏览窗口。编辑程序前先设置当前工作文件夹。单击 jGRASP 界面左侧的工作文件夹下拉列表，如图 1-16 所示工具栏下方的下拉列表。

图 1-16　工作文件夹

在下拉列表中展开子文件夹或者单击↑工具按钮打开父文件夹选择项目。比如设置工作文件夹为 code_1_2。双击文件夹中的 .java 文件或者使用菜单命令 File|Open 即可在编辑器中打开 Java 源文件。使用菜单命令 File|New 新建 Java 类。比如新建 Calculator 类，jGRASP 以默认文件名[jGrasp♯♯]作为源文件名，打开编辑窗口，进入编辑状态。使用菜单命令 View|Line Numbers 在编辑窗口中显示行号，或者单击工具按钮显示行号，如图 1-17 所示。按照源文件 1-3 所示源代码录入完毕后，单击工具栏中的"保存"按钮保存。jGRASP 打开"另存为"对话框，如图 1-18 所示。保持默认设置，使用默认的文件名，单击 Save 按钮保存。

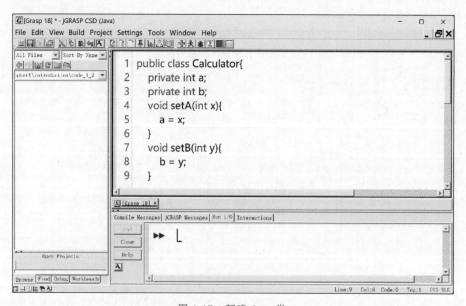

图 1-17　新建 Java 类

单击工具栏中的 + 按钮编译光标插入点所在的源文件。如果存在编译错误，比如在第 5 行缺少分号，那么 jGRASP 会在 Compile Messages 窗格中输出编译消息：Calculator.java:5: error: ';' expected，如图 1-19 所示。把光标移动到第 5 行末尾，插入西文分号，重新

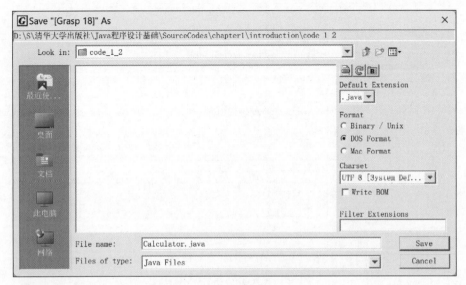

图 1-18 "另存为"对话框

编译。如果没有任何编译错误消息,表示编译成功。

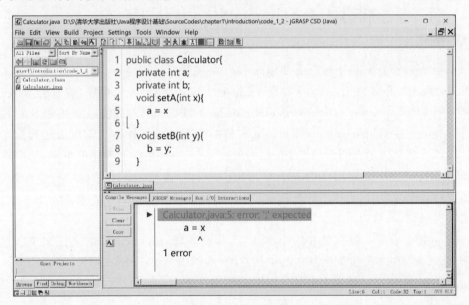

图 1-19 编译消息

Calculator.java 完成编译后,再新建 Test.java 来创建和使用 Calculator 对象。再次从 File 菜单中选择 New|Java 新建 Java 类,按照默认的文件名保存,然后编译。

此时通过编译在工作文件夹中得到两个字节码文件 Calculator.class 和 Test.class。由于程序的入口点 main 方法在 Test 类中,所以把光标放置在 Test.java 的编辑窗口中,单击 按钮,从 Test 类的 main 方法开始运行。运行结果显示在 Run I/O 窗格中,如图 1-20 所示。

如果程序的输出结果与预期结果不同或运行期间抛出异常,则需要单击 按钮进入调试状态确定出错位置,改正错误。附录中有关于 jGRASP 更为详尽的操作介绍。

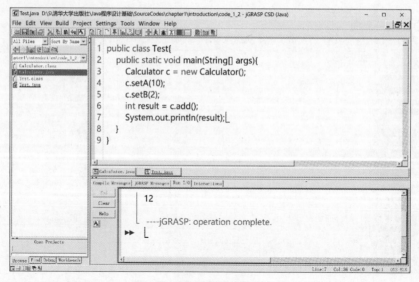

图 1-20 运行

1.5 使用 Eclipse 设计程序

Eclipse 是在生产环境中被广泛使用的 Java 集成开发环境之一。Eclipse 本身也由 Java 语言开发，所以为了使 Eclipse 运行和在 Eclipse 中编译运行 Java 程序，应首先安装 JDK。然后从官网 https://www.eclipse.org/downloads/下载 Eclipse，目前最新的版本是 Eclipse 2023。下载的文件 eclipse-inst-jre-win64.exe 是一个可执行文件。只需双击运行这个可执行文件即可安装。为了方便启动 Eclipse，一般在 Windows 桌面上创建 Eclipse 快捷方式。

Eclipse 把开发一个应用作为一个"项目（project）"管理。一个项目中有多个源文件。当类的数目越来越多时，为了方便管理，就需要把相关的类组织在一起，称为放在"包"中；并通过名字进行标识，称为包名。

Eclipse 把所有的项目存储在一个称为"工作空间（work space）"的文件夹中。第一次运行 Eclipse 时，Eclipse 会询问默认的工作空间（本例中为 C:\SEWORKSPACE）。

在 Windows 桌面上双击 Eclipse 快捷方式，启动 Eclipse 后设计程序的步骤一般是首先在 Eclipse 中新建项目，然后新建类，最后运行。下面以编辑运行源文件 1-3 和源文件 1-4 为例介绍 Eclipse 环境的使用。

1. 新建 Java 项目

启动 Eclipse，从 File 菜单中选择 New|Java Project。在弹出的 New Java Project 对话框中输入项目名称（Project Name）：examples。

接着设置项目位置（Project Location）：一般勾选使用默认的位置（Use default location）。Eclipse 就把项目创建在默认工作空间的与项目名同名的文件夹中，如图 1-21 所示。

其他设置保持不变，单击 Finish 按钮。项目只需创建一次。再次运行 Eclipse，Eclipse 会自动打开这个项目。

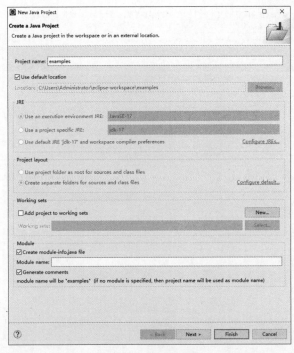

图 1-21 创建项目

2. 新建 Java 类

在包浏览器(Package Explorer)窗格中右击项目名称 examples，在弹出的快捷菜单中选择 New|Class。在弹出的 New Java Class 对话框中键入类名(Name)：Calculator，如图 1-22 所示。

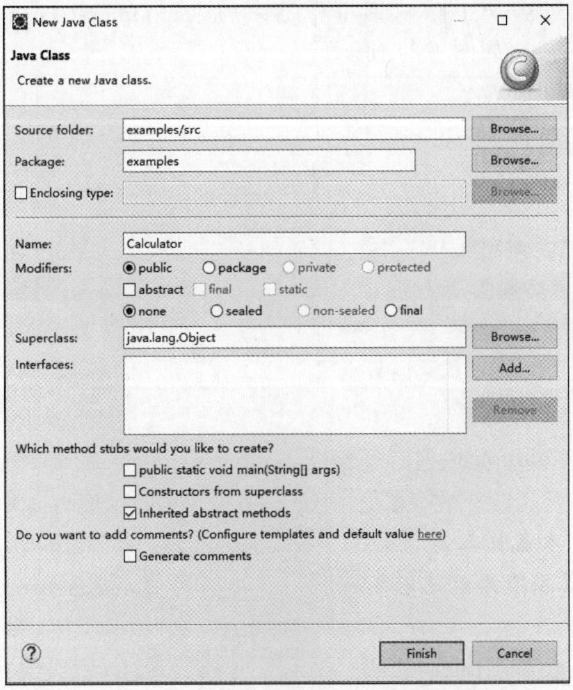

图 1-22 New Java Class 对话框

单击Finish按钮,在编辑器窗口中录入源文件1-3,如图1-23所示。

图1-23 编辑器窗口

第1行使用保留字package把类放到包examples中。第3行使用保留字class定义类。类的名字就是刚在对话框中键入的Calculator。保留字public称为访问修饰符,声明该类可以被其他任意类使用。单击"保存"快捷按钮。

然后去创建第二个类:Test。再次右击项目名称,在弹出的快捷菜单中选择New|Class。在弹出的New Java Class对话框中键入类名(name):Test,并勾选public static void main(String[] args),如图1-24所示。

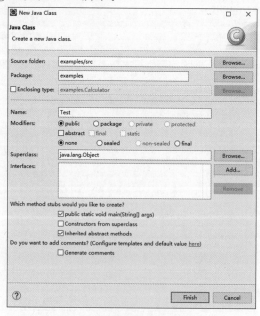

图1-24 新建含main方法的类

在编辑窗口中录入源文件 1-4,如图 1-25 所示。

图 1-25　编辑窗口中的 Test.java

单击"保存"按钮。

3. 运行 Java 程序

最后运行程序。在包资源浏览器中右击 Test 类,在弹出的快捷菜单中选择 Run As|Java Application。在 Console 窗口中观察运行结果,如图 1-26 底部所示。

图 1-26　在 Console 窗口中观察运行结果

一般地，一个 Java 项目中含有若干包；一个 Java 包中含有若干类。一个类是由若干成员组成的，这些成员包括变量和方法。Java 程序只能从约定的 main 方法开始执行。

如果需要，Java 程序可以脱离 Eclipse 环境运行。在包浏览器中右击项目名称，在弹出的快捷菜单中选择 Export，弹出 Export 对话框，如图 1-27 所示。

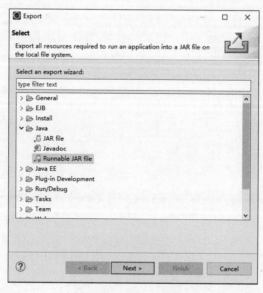

图 1-27　Export 对话框

在 Export 对话框中选择 Runnable JAR file 作为输出目标（export destination），如图 1-27 所示。

单击 Next 按钮。从启动配置（Launch configuration）下拉列表中选择 Test-examples；单击 Browse…按钮，设置输出目标（Export destination）为 C:\work\Test.jar，如图 1-28 所示，单击 Finish 按钮。

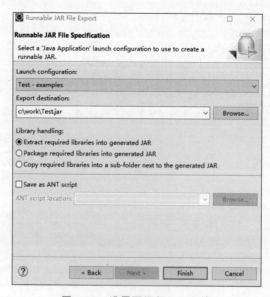

图 1-28　设置可运行 Jar 文件

最后打开命令提示窗口，查看和使用 java -jar 运行所生成的 jar 文件，如图 1-29 所示。

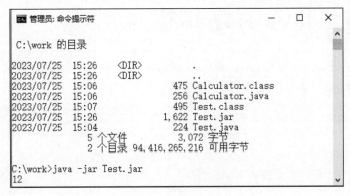

图 1-29　运行 jar 文件

由于 Test 类使用了 Calculator 类，所以称 Test 类是 Calculator 类的客户。Test.java 也就是 Calculator.java 的客户程序。

如果包浏览器窗口由于误操作而消失了，那么从 Window 菜单中选择 Show View | Project Explore 重新显示包浏览器。

假设在第一次启动 Eclipse 时设置的默认工作空间是 D:\SEWORKSPACE，并假设在此工作空间中已经完成了项目 examples，现需要把工作空间复制到另外一个位置，比如 E:\，那么：

（1）关闭 Eclipse，在 Windows 资源管理器中选择 D:\SEWORKSPACE 文件夹，按下 Ctrl＋C 组合键进行复制。

（2）在 Windows 资源管理器中打开 E 盘，按下 Ctrl＋V 组合键进行粘贴。

（3）再次启动 Eclipse，默认打开了工作空间 D:\SEWORKSPACE。从 File 菜单中选择 Switch Workspace | Other，在弹出的 Workspace Launcher 对话框（图 1-30）中单击 Browse 按钮，选择 E:\SEWORKSPACE，单击 OK 按钮，那么就完成了工作空间的切换。

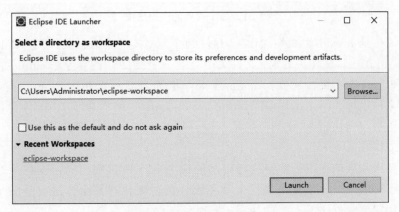

图 1-30　切换工作空间

使用这个办法可以从具有硬盘还原卡的实验室计算机上备份工作空间到 U 盘，然后再次上机时，把 U 盘中的备份复制到实验室计算机上，再启动 Eclipse 并把默认的工作空间切换到复制的工作空间上。

1.6 Java 程序结构

一个 Java 应用由很多类组成，这些类按照相关性组织在不同的包中。图 1-31 展示了 Java 程序的组织结构。

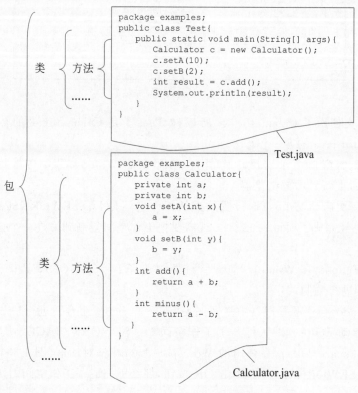

图 1-31　Java 程序的组织结构

包 Package 是 Java 中组织类的设施，Java 每个包对应一个磁盘文件夹。本例中包的名字为 examples，在项目文件夹的 src 文件夹中就有一个文件夹名字为 examples。包中的类 Calculator 和 Test 对应的源文件 Calculator.java 和 Test.java 保存在 examples 文件夹中。

一个源文件也称为一个编译单元（compilation unit）。编译命令 javac 负责检查源程序中是否存在语法与词法错误，比如语句是否以分号结束等。编译器识别的错误称为编译时刻错误（compile-time error）。

由负责装入和启动虚拟机解释执行程序的命令 java 所识别出的错误称为运行时刻错误（run-time error）。比如程序如果让一个不存在的对象执行某个方法就是一个运行时刻错误。Java 以异常（exception）机制处理运行时刻错误。

如果程序在编译时刻和运行时刻均没有发生错误，但是运行结果并不正确，则程序中存在逻辑错误（logical errors）。比如，问题需要对 10 个整数进行求和，但程序只累加了 9 个数，这类错误就是逻辑错误。一般通过调试（debugging）来定位和解决逻辑错误。

1.7 代码风格

良好的代码风格(code style)有益于提高程序的可读性,使 Java 程序员能够更快更轻松地理解源代码。代码风格包括命名约定、留白以及块风格等。

1.7.1 命名约定

Java 中标识符的命名一般遵从以下约定:

(1) 只有一个单词的类名首字母大写。如果类名中包含多个单词,则每个单词的首字母大写。这称为大驼峰命名风格。

(2) 只有一个单词的变量名和方法名中的字母均小写。如果变量名或方法名中含有多个单词,则除了第一个单词外,其他单词首字母大写。这称为小驼峰命名风格。

(3) 常量名的所有字母大写。如果名字中含有多个单词,则使用下画线隔开,如 MAX_VALUE。

注意:一定选择能够体现类、变量和方法含义的英文单词作为其标识符,不要使用汉字或者汉语拼音作为标识符。

1.7.2 留白

"留白"是绘画中的术语,指在一幅画中要有空白。源代码中也要有空白,以形成清晰的逻辑结构。虽然留白不会影响程序的正确性,但影响程序的可读性。缩进、空行和空格是三种留白的方法。缩进(Indentation)使用平面空间的布局关系反映了程序部件间的逻辑关系(比如包含关系)。观察如下代码:

```
public class Calculator{
    private int a;
    private int b;

    void setA(int x){
        a = x;
    }

    void setB(int y){
        b = y;
    }

    int add(){
        return a + b;
    }

    int minus(){
        return a - b;
    }
}
```

在类 Calculator 中定义 6 个成员:2 个变量和 4 个方法。每个成员均缩进 4 个空格以反映这种包含关系。方法中的语句也缩进4个空格，同样反映了这些语句与其所属方法间的包含关系。禁止使用制表符(\t)实现缩进，因为在不同的软件中对制表符的显示结果不同。

每个二元运算符的前后都有一个空格(blank)。例如：

```
a + b;          //好的风格
a+b;            //不好的风格
```

方法参数在定义和传入时，多个参数逗号后边必须加空格。
不同代码片段应使用空行隔开。例如类的定义、方法的定义前有一空行。
空格、制表符、回车、换行、进纸等都是空白符(whitespace)，编译器忽略空白符。

1.7.3 块风格

块是使用一对花括号"{}"括起来的一组语句。有两种块风格：next-line风格和end-of-line风格。

next-line风格为块另起一行并把开括号"{"放在新行的行首，例如：

```
class A
{
    void aMethod()
    {
        //Do something
    }
}
```

end-of-line风格不为块另起一行，而是直接把开括号"{"置于行尾，并在前面有一个空格，例如：

```
class A {
    void aMethod() {
        //Do something
    }
}
```

1.8 注释

Java源代码注释有两种：实现注释(implementation comments)和文档注释(documentation comments)。实现注释分两种：多行注释和单行注释。

多行注释形如 /*...*/，即从 /* 开始到 */ 结束。例如：

```
/*
 * This is a multiple line comment.
 */
```

一般约定 * 和后面的注释文本间有一个空格，所有行的 * 垂直对齐。
单行注释形如//…，即从//开始到行尾结束。例如：

```
//This is the first line of a multi-line comment.
//This is the second line of a multi-line comment.
//This is the third line of a multi-line comment.
```

一般约定双斜线//与注释文本之间有一个空格。单行注释和多行注释一般出现在方法体中在被注释语句的上方，并与被注释代码左对齐。

文档注释供文档抽取工具 javadoc 使用，形如/＊＊…＊/，例如：

```
/**
 * The basic Car class
 * @author Donald . Dong (original)
 */
```

与多行注释一样，一般约定＊和后面的注释文本间有一个空格，所有行的＊垂直对齐。文档注释中的注解有以下 4 类：

用于包：@see,@link,@since,@deprecated；

用于类和接口：@see,@link,@since,@deprecated,@author,@version；

用于成员变量：@see,@link,@since,@depreciated,@serial,@serialfield；

用于成员方法：@see,@link,@since,@deprecated,@param,@return,@throws,@serialdata。

@author 表示后面是类或接口的创建者，例如@author Donald Dong。@deprecated 表示该项目已经被新版本替换，不建议再使用。@return 表示后面的文本是对方法的返回值的解释。@param 表示其后面是参数名字以及含义。@throws 表示其后面是抛出的异常类型以及描述。@see 表示"参阅"，其后面是对其他文档的引用。@since 表示后面是实体的起始版本，例如@since 1.2 或者是起始日期，例如@since 20 Jan. 2023。@Override 表示本方法是覆盖方法。@link 表示后面是对相关代码的超链接。@code 表示后面是 Java 源代码中的词元，比如类名或者方法名。

类、成员方法、成员变量的注释必须使用文档注释，即/＊＊注释文本＊/格式，不能使用单行注释。所有的类都必须有创建者和创建日期；所有方法头部的上方除了注释返回值、参数、异常说明外，还必须指出该方法做什么事情，实现什么功能。所有的覆盖方法必须加@Override 注解。

Javadoc 并不保留文档注释中的换行和多余的空格。为了按照期望的格式生成 API 文档，则应使用 HTML 标记设计输出格式。比如：
换行，<p>另起一段，<pre></pre>保留其中的换行和多余的空格等。

下面是 JDK String 类的文档注释（有删减）：

```
/**
 * The {@code String} class represents character strings. All
 * string literals in Java programs, such as {@code "abc"}, are
 * implemented as instances of this class.
 * <p>
 * Strings are constant; their values cannot be changed after they
 * are created. String buffers support mutable strings.
 * Because String objects are immutable they can be shared. For example:
 * <blockquote><pre>
 *     String str = "abc";
```

```
 * </pre></blockquote><p>
 * is equivalent to:
 * <blockquote><pre>
 *     char data[] = {'a', 'b', 'c'};
 *     String str = new String(data);
 * </pre></blockquote><p>
 * Here are some more examples of how strings can be used:
 * <blockquote><pre>
 *     System.out.println("abc");
 *     String cde = "cde";
 *     System.out.println("abc" + cde);
 *     String c = "abc".substring(2,3);
 *     String d = cde.substring(1, 2);
 * </pre></blockquote>
……
 * @author  Lee Boynton
 * @author  Arthur van Hoff
 * @author  Martin Buchholz
 * @author  Ulf Zibis
 * @see     java.lang.Object#toString()
 * @see     java.lang.StringBuffer
 * @see     java.lang.StringBuilder
 * @see     java.nio.charset.Charset
 * @since   1.0
 */
public final class String
    implements java.io.Serializable, Comparable<String>, CharSequence {
}
```

从源代码中把以上文档注释抽取出来形成 HTML 帮助文档,在浏览器中显示为:

The String class represents character strings. All string literals in Java programs, such as "abc", are implemented as instances of this class.

Strings are constant; their values cannot be changed after they are created. String buffers support mutable strings. Because String objects are immutable they can be shared. For example:

```
String str = "abc";
```

is equivalent to:

```
char data[] = {'a', 'b', 'c'};
String str = new String(data);
```

Here are some more examples of how strings can be used:

```
System.out.println("abc");
String cde = "cde";
System.out.println("abc" + cde);
String c = "abc".substring(2, 3);
String d = cde.substring(1, 2);
……
```

下面是 String 类中 charAt 方法的文档注释：

```
/**
 * Returns the {@code char} value at the
 * specified index. An index ranges from {@code 0} to
 * {@code length() - 1}. The first {@code char} value of the sequence
 * is at index {@code 0}, the next at index {@code 1},
 * and so on, as for array indexing.
 *
 * <p>If the {@code char} value specified by the index is a
 * <a href="Character.html#unicode">surrogate</a>, the surrogate
 * value is returned.
 *
 * @param      index   the index of the {@code char} value.
 * @return     the {@code char} value at the specified index of this string.
 *             The first {@code char} value is at index {@code 0}.
 * @exception  IndexOutOfBoundsException  if the {@code index}
 *             argument is negative or not less than the length of this
 *             string.
 */
public char charAt(int index) {}
```

那么，其抽取出的帮助文档在浏览器中显示为：

Returns the char value at the specified index. An index ranges from 0 to length() - 1. The first char value of the sequence is at index 0，the next at index 1，and so on，as for array indexing.

If the char value specified by the index is a surrogate，the surrogate value is returned.

Parameters：

　　index - the index of the char value.

Returns：

　　the char value at the specified index of this string. The first char value is at index 0.

Throws：

　　IndexOutOfBoundsException- if the index argument is negative or not less than the length of this string.

变量和常量上的文档注释主要内容是其设计意图、用途等。例如：

```
/**
 * The value is used for character storage.
 */
private final byte[] value;
```

第 1 章　章节测验

第2章 Java语言基础

本章讨论 Java 语言中的标识符、保留字、数据类型、字面量、变量、运算符、表达式、控制结构以及数组。

2.1 标识符和保留字

类的名字、变量的名字、方法的名字等统称为标识符。标识符要符合以下规则。

(1) 标识符必须以字母(包括 Unicode 字母)、下画线(_)或美元符号($)开始。

(2) 除了第一个字符外,其余部分由字母、数字、下画线或美元符号组成。

(3) 不能使用 Java 保留字作为标识符,如 class、public、void 等。

(4) 标识符中不能含有特殊符号,如叹号(!)、at号(@)、数字符号(♯)、百分号(%)、与号(&)、尖号(^)、星号(*)或者空格。

有效的标识符如 Car、MyCar、length、licenceNumber、_START、$ END 等;无效的标识符如 123car、my-car、class(class 是一个 Java 保留字)等。Java 标识符是大小写敏感的。例如,标识符 Calculator 和 calculator 是两个不同的标识符。建议使用与实际意义一致的英文单词作为标识符以提高代码可读性,并遵循驼峰命名法等约定。例如,对于类名使用大驼峰命名法(每个单词首字母大写,如 MyCar),而对于变量名则采用小驼峰命名法(首个单词首字母小写,后续单词首字母大写,如 licenceNumber)。

保留字(Reserved keyword)是 Java 语言中保留使用的,具有特定含义的标识符,如 class、public、void 等。下面以字母顺序列出了 Java 21 种的保留字:

abstract	assert	boolean	break	byte	case
catch	char	class	const	continue	default
do	double	else	enum	extends	final
finally	float	for	if	goto	implements

import	instanceof	int	interface	long	native
new	package	private	protected	public	return
short	static	strictfp	super	switch	synchronized
this	throw	throws	transient	try	void
volatile	while	_ (underscore)			

在 Java 中,null 不是保留字,而是一个字面量(literal),表示"没有",即某个引用类型的变量没有引用任何对象。true 和 false 也不是保留字,而是布尔类型的字面量,分别标识布尔值:真(true)和假(false)。

2.2 基本数据类型

数据类型(data type)是对值以及在这些值上运算的约束。Java 中有两种类型:引用类型(reference types)和基本数据类型(primitive types)。

引用类型包括 5 种:类(class)、接口(interface)、数组(array)、枚举(enum)和注解(annotation)。所谓"引用",就是用来访问对象的标识。无论是身份证上正式登记的姓名,还是自定义的"网名",都是用来"引用"一个人的,此时的引用是一个字符串;可以通过订单号来引用某个订单,此时的引用就是一个数字串。创建一个对象就在内存中分配了一块空间来存储该对象,可以使用对象的类型和对象内存空间的起始地址来引用某个对象,例如类 Calculator 的某对象的引用形如 Calculator@24d46ca6,意思是"在起始地址 24d46ca6 处的 Calculator 对象"。

基本数据类型有 8 种:boolean、char、byte、short、int、long、double 和 float。

数据类型 boolean 表示逻辑类型,逻辑值只有两个:true 和 false。例如语句:boolean result = true;定义了名字为 result 的变量,该变量是 boolean 类型,并且赋值给该变量的值为 true。对该类型变量的操作有与、或、非等。

数据类型 char 表示字符类型,定义了 Unicode 字符集及其上的操作。字符集就是一个按照特定顺序排列的字符集合。Unicode 字符集包含了世界上很多语言中的字符和符号。附录 A 就是该字符集中的一部分(Unicode Chart - Basic Latin)。源程序中的字符要使用单引号(')括起来。例如:

```
'a'        //字母 a
'\t'       //制表符
```

也可以使用 Unicode 代码点来表示字符,如'\u0061' 就是 'a'。在'\u0061'中,0061 是 4 位的十六进制数字,表示英文字母"a"在 Unicode 字符集中的编号;"\u"表示把"0061"转义为 1 个 Unicode 字符而不是 4 个普通十六进制数字。

Java 语言已经使用了单引号标记单个字符,即单引号有了特殊用途。如果在源程序中把单引号作为普通字符使用,则需要使用转义符号"\"。也就是说,\'表示普通单引号字符。

字符串是字符序列。Java 中使用类型 String 表示字符串,但是它不是基本数据类型而是引用类型"类"。

Java 中把数(Number)分成两类:整数和浮点数。所有数都是有符号数,也就是说,都有正数和负数。整数有4种类型：byte、short、int 和 long。这4种类型的区别在于用于表示这些数值的比特数不同，如表2-1所示。整数上的运算有加、减、乘、除和取余等。

表 2-1 整数类型

类型	大小/b	最小值	最大值
byte	8	−128	127
short	16	−32 768	32 767
int	32	−2 147 483 648	2 147 483 647
long	64	-2^{63}	$2^{63}-1$

浮点类型包括 float 和 double。浮点数的表示形式为：

$$<有效数字> \times <底>^{<指数>}$$

在这种表示中，小数点的位置取决于"指数"，即通过"指数"可以把小数点的位置"浮动"到有效数字部分的任意位置。算术运算符 +、−、*、/ 均可用于浮点数运算。Java 语言中的 Math 类提供了平方根(square root)、对数(logarithm function)、幂(exponential function)等运算。浮点类型 float 使用 32 比特(4 字节)，其有效数字占用 24 比特(约 7 个十进制数字)；而浮点类型 double 使用 64 比特(8 字节)，其有效数字部分占据了 53 比特(约 16 个十进制数字)。浮点数的"精度"指有效数字的位数。

注意，浮点数不能精确地表示所有的实数。比如十进制数 0.1 只能被近似表示为二进制数 0.000110011001100110011010。float 类型的 0.1 与实数 0.1 的误差为 2^{-25}。

2.3 字面量

字面量(literal)是值在源代码中的文本表示。根据数据类型不同，Java 中字面量有布尔类型字面量、字符类型字面量、整数类型字面量、浮点类型字面量和字符串字面量(String 类型)。

布尔类型字面量只有两个：true 和 false。true 表示逻辑值"真"；false 表示逻辑值"假"。

字符类型字面量就是单引号引起来的单个字符。Java 将其按 16 位 Unicode 字符编译。Unicode 字符集把一个字符映射到一个整数，称为 Unicode 代码点，基本平面中的代码点最小是 0，最大是 65 536，使用十六进制表示就是从 0000 到 FFFF。所以 Java 中可以使用 Unicode 转义'\u0000'到'\uFFFF' 来表示单个字符，其中"u"的意思就是 Unicode。例如：

```
char ch = 'a';              //a
char familyName = '张';      //张
char ch = '\u6B22';         //欢
```

如果源代码编辑器不支持显示中文，则只好使用'\u6B22'；如果编辑器支持中文，当然应该直接写'欢'。

诸如换行之类的特殊字符，则使用转义符号(\)。例如，'\n' 表示换行符。其他特殊符号如表 2-2 所示。

表 2-2　特殊转义序列

记号	字符	Unicode 转义表示	记号	字符	Unicode 转义表示
\n	换行	\u000a	\t	制表	\u0009
\r	回车	\u000d	\"	双引号	\u0022
\f	进纸	\u000c	\'	单引号	\u0027
\b	退格	\u0008	\\	反斜线	\u005c
\s	空格	\u0020			

字符串字面量是使用双引号引起了的字符序列,例如 "Java"。注意,无论单引号还是双引号,都是英文引号而不是中文引号。

根据进制不同,整数字面量分为十进制整数、十六进制整数和二进制整数。默认使用十进制整数,例如 12。如果想使用十六进制整数,则需要在数字序列前面添加"0x"或者"0X"前缀。例如 0xC 表示十进制数 12。Java SE 7 及后续版本支持二进制整数字面量,如 0b101。二进制整数字面量则需要在 01 数字序列前面添加"0b"或者"0B"前缀。下面的语句演示了不同进制的整数字面量:

```
byte b = 0x7f;          //127,十六进制
int x = 0b101;          //5,二进制
int i = 0x1f;           //31,十六进制
int j = 0X1F;           //31,十六进制
int k = 31;             //31,十进制
long m = 200L;          //200,十进制
long n = 200l;          //200,十进制
```

Java 默认所有整数字面量的数据类型为 int。如果指定整数字面量为 long 类型,则需要在数字的末尾添加后缀"L"或者"l"。例如 2L 表示整数 2 为 long 类型。注意不能把"宽"的字面量赋值给"窄"的变量,例如 byte b＝0x7ff,0x7ff 是 32 位整数,赋值给 8 位的变量,编译会提示错误。但是编译器并不对 byte b＝0x7f;提示错误,这是因为编译器识别到虽然 0x7f 是 32 位整数,但高 24 位都是 0,截断后没有损失。

浮点数字面量有两种表示形式:标准形式和科学记数形式。例如 583.45 是标准形式,1.23e＋2 是科学记数形式。字面量 1.23e＋2 表示数字 $1.23\times 10^2＝123$。注意,1.23e＋2 与 1.23e2 含义相同。"e"表示指数(exponent),也可以用大写"E"。

Java 默认浮点字面量是 double 类型。如果要指定为 float 类型,则需要在数字的末尾添加后缀"F"或者"f"。可以使用后缀"D"或者"d"显式声明为 double 类型的字面量。例如:

```
float f1 = 128.6F;           //32 位浮点数,标准形式
float f2 = 128.6f;           //32 位浮点数,标准形式
float f3 = 1e-45f;           //10⁻⁴⁵,科学记数形式
float f4 = 1e+9f;            //10⁹,科学记数形式
double d1 = 1256.8d;         //64 位浮点数,标准形式
double d2 = 1256.8D;         //64 位浮点数,标准形式
double d3 = 1.2568e3d;       //1.2568×10³,科学记数形式
```

2.4 变量

变量(variable)是一个命名的存储单元。该存储单元中的内容在程序运行期间可能会被改变。变量的数据类型决定了存储单元的大小。变量必须先声明其数据类型才能使用。Java 是强类型语言，这是因为 Java 先检查类型一致性才去运行程序。例如，源程序中的表达式"2＋true"，Java 就会发现这个类型不一致的错误，因为一个整数和一个布尔值相加的运算符"＋"在语言中没有定义。

变量有 4 种类型：实例变量(Instance Variables)、类变量(Class Variables)、局部变量(Local Variables)和形式参数(Parameters)。实例变量是每个对象都有的成员变量。实例变量随着对象的创建而存在，随着对象的消失而消失。创建对象时如果没有给实例变量显式赋值，Java 会赋予默认初值。类变量仅仅属于类而不属于对象。声明成员变量时如果使用保留字 static 修饰，则是类变量；否则就是实例变量。在方法体中声明的变量称为局部变量。方法每被调用一次就会为方法体中的局部变量重新分配一次存储；方法执行完毕，局部变量的存储空间被回收。只能在声明局部变量的方法体中访问局部变量。局部变量的初值必须由程序员显式指定。方法的形式参数也是局部变量。后继课程"编译原理"会介绍如何实现方法调用、方法返回和参数传递。

在类体或者方法体的任意位置均可以声明变量，只要在变量使用之前进行声明即可。声明变量的语法：

```
<数据类型> <变量名> [=<初值>];
```

声明变量时既可以指定初值，也可以不指定。如果声明变量并初始化，那么使用语法：

```
<数据类型><变量名> = <初值>;
```

如果先声明再初始化，那么使用语法：

```
<数据类型> <变量名>;
<变量名> = <初值>;
```

变量的命名要服从 Java 的标识符规则，为了程序可读性，要求变量的名字要与实际意义一致。比如一个表示年龄的变量应使用 age 而不是使用单个字符 a。

下面是声明变量的几个例子。

```
int x;                  //声明整数类型的变量,名字为 x
double radius;          //声明浮点数类型变量,名字为 radius
char a;                 //声明字符类型变量,名字为 a
double grade = 0.0;     //声明浮点类型变量,名字为 grade,初值为 0.0
```

为了在屏幕上显示变量的值，可以使用 System.out.println()或者 System.out.print()。System.out.println()完成输出后自动换行而 System.out.print()不会自动换行。System 是 JDK 中的类，out 是 System 类中声明的静态成员变量，通过该变量引用了一个负责进行输出的对象。println()是 out 对象的实例方法。System.out.println("Hello")的意思是：执行 System.out 对象上的方法 println()输出字符串"Hello"。又比如下面的程序片段

```
int value = 10;
char x;
x = 'A';
System.out.println(value);
System.out.println("The value of x= " + x);
```

的输出为：

```
10
The value of x = A
```

System.out.println("The value of x＝"＋x);中的"＋"进行字符串连接。在连接之前，首先会把非字符串类型的变量 x 转换为字符串类型。

多个变量可以一起直接声明并初始化。例如：

```
int x, y;
int a = 3, b = 4, c = 5;
```

基本数据类型的变量把数据存放在该变量名标识的存储单元中；而引用类型的变量仅仅在该变量名标识的存储单元中存放所引用的对象的地址，即指向了存放对象的存储单元。

假设有如下声明：

```
int i = 10;                    //基本数据类型
String name = "Donald";        //引用数据类型
```

表 2-3 显示了这两个变量的变量名字、存储单元地址和存储单元中的值之间的关系。

表 2-3 变量的地址、名字和值

存储单元地址	变量名字	存储单元中的值
0X08100000	i	10
...		...
0X081000F0	name	0X0810A000
...		...
0X0810A000		"Donald"

在进行表达式计算时，Java 能够自动地把"窄"数据类型转换为"宽"数据类型。所谓"窄"，就是该数据类型所用的存储字节少；所谓"宽"，就是该数据类型所用的存储字节多。自动转换规则如下：

byte→short→int→long→float→double

char→int→long→float→double

例如，如果程序中有表达式 2 * 0.3，那么就会把 2 转换为 double 类型，然后再和 double 类型的字面量 0.3 相乘。

注意：boolean 类型不参与转换。byte、short 和 char 这三种基本类型如果参与数学运算，那么先自动提升成 int 类型，再参与数学运算；如果不参与数学运算，则按照基本类型的自动转换规则进行转换。

反过来，把"宽"数据类型转换为"窄"数据类型则可能会由于部分数据位被截断而产生错误。如果程序员确信不会产生错误，或者愿意承担产生错误的风险，则可以通过显式"转型(cast)"实现"窄化"。在变量前面添加括号括起来的数据类型就能把该变量转型为此数据类型。例如,下面的赋值语句就把 long 类型的变量 g 赋值给了 int 类型的 i:

```
int i;
long g = 100;
i = (int) g;
```

转型运算(int)把 long 值转换成 int 值。

源文件 2-1 演示了直接把 double 类型的 j 赋值给 int 类型的 i 则会产生编译错误(第 5 行);而显式的转型(cast)则没有错误(第 6 行)。

源文件 2-1   CastingDemo.java

```
1 class CastingDemo {
2     public static void main(String[] argv) {
3         int i;
4         double j = 2.75;
5         i = j;                        //编译错误
6         i = (int) j;                  //显式转型;i 的值为 2
7         System.out.printf("%s%d\n", " i = ", i);
8         long b;
9         b = i;                        //没有编译错误
10        System.out.printf("%s%d\n",  " b = ", b);
11    }
12 }
```

程序的输出为

```
i = 2
b = 2
```

在对 byte、short 和 char 这三种数据类型变量使用字面量初始化时,如果初值在其数据类型的取值范围内,编译器自动优化为该数据类型。例如,byte b=0x2A;仍然会把值 42 存入变量 b 而不会出现编译错误,虽然字面量 0x2A 是一个整数;而在 byte b=0x8A;中 0x8A 是正整数 138,超出了 byte 类型的范围[-128,127],所以会出现编译错误。

2.5 运算符

运算符是组成表达式的重要成员。Java 语言中的运算符有赋值运算符、算术运算符、关系运算符、逻辑运算符、条件运算符等。如果一个运算符有两个操作数,则称其为双目运算符(binary operator);如果只有一个操作数,则称其为单目运算符(unary operator)。表达式中的运算符服从一定的优先级规则。

2.5.1 赋值运算符

赋值(assignment)运算符用以向变量名标识的存储单元中存放一个值。下面是几个赋值语句:

```
//把整数 2 赋值给变量 a
int a = 2;

//把表达式 2 + 3 的运算结果赋值给 b
int b = 2 + 3;

//把字符串 " Donald " 的引用赋值给引用类型变量 name
String name = "Donald";

//把 b 赋值给 a,然后把 a 赋值给 d
int d = a = b;
```

在赋值语句中,首先对赋值运算符(=)右侧的表达式计值,然后把计值结果存放到左侧变量名标识的存储单元中。一个变量只能存放与其类型兼容的一个值。所谓"兼容",指值的数据类型能够自动转型到该变量的数据类型。如果再次对变量赋值,那么新的值将替换原来的值。

2.5.2 算术运算符

算术(arithmetic)运算符包括＋、－、＊、/和 ％。其含义如表 2-4 所示。

表 2-4 算术运算符

表达式	含 义	表达式	含 义
a＋b	a 加 b	a/b	a 除以 b
a－b	a 减 b	a％b	a 除以 b 的余数
a＊b	a 乘以 b		

如果 a 和 b 都是整数,那么除运算/只保留商而舍弃余数。如果 a 和 b 其中有一个是浮点数,则执行浮点数的除法运算。例如,5/2 的结果是 2,而 5.0/2、5/2.0、5.0/2.0 的结果都是 2.5。

取余运算符％也称为模运算符,其结果是整数除法运算 a/b 的余数。取余运算 a％b 结果的符号取决于第一个操作数 a。例如:－10％3 的结果是－1;而 10％－3 的结果是 1。

当算术运算符＋和－用作单目运算符时,就表示正数和负数。运算符＋＋和－－也是单目运算符。增量运算符(＋＋)把操作数增 1 而减量运算符(－－)从其操作数中减去 1。如果＋＋和－－写在变量的前面,则称为该运算符的前缀形式;如果＋＋和－－写在变量的后面,则称为该运算符的后缀形式。后缀形式和前缀形式表达式的计值结果不同:后缀形式把增 1 前的值作为表达式的值;而前缀形式则把增 1 后的值作为表达式的值。例如,假设变量 i 的当前值是 1,下面的语句将使 j 的值为 1 而 i 的变为 2:

```
j = i ++;
```

但是,下面语句使 i 和 j 的值为 2:

```
j = ++ i;
```

当运算符＋的两个操作数中存在字符串时则进行字符串连接运算;而当操作数中存在字符时则进行整数加法运算,如源文件 2-2 所示。

源文件 2-2　OperatorAdditionDemo.java

```
1 public class OperatorAdditionDemo {
2     public static void main(String[] args) {
3         System.out.println("ABC" + 123);
4         System.out.println(12 + 3 + "abc");
5
6         System.out.println('A' + 2);
7         System.out.println("B" + 2);
8         System.out.println("B" + 'A' + 2);
9     }
10 }
```

程序的输出为：

```
ABC123
15abc
67
B2
BA2
```

因为字符 A 的 Unicode 代码点是\u0041,整数值为 65,65＋2 等于 67,所以第 6 行的输出是 67。

2.5.3　关系运算符

关系运算符(relational operator)计算两个操作数的大小关系。其运算结果为 true 或 false。例如大于、大于或等于、相等、不相等。Java 中的关系运算符有：＝＝（is equal to）、！＝(is not equal to)、＜(is less than)、＞(is greater than)、＜＝(is less than or equal to)和＞＝(is greater than or equal to)。这些运算符的含义如表 2-5 所示。

表 2-5　关系运算符

表达式	含义	表达式	含义
a＞b	a 大于 b?	a＜＝b	a 小于或等于 b?
a＞＝b	a 大于或等于 b?	a＝＝b	a 等于 b?
a＜b	a 小于 b?	a!＝b	a 不等于 b?

例如下面代码片段输出 false,因为 2 和 3 不相等。

```
int a = 2;
int b = 3;
System.out.println(a == b); //返回 false
```

注意：＝＝是关系运算符,而＝是赋值运算符。

源文件 2-3 演示了关系运算符的使用。

源文件 2-3　OperatorComparisonDemo.java

```
1 public class OperatorComparisonDemo {
2     public static void main(String[] args) {
3         int a = 3;
4         int b = 4;
5         int c = 5;
6
7         System.out.print((a < b) + ", ");
```

```
 8      System.out.print((a <= b) + ", ");
 9      System.out.print((b < c) + ", ");
10      System.out.print((b <= c) + ", ");
11
12      System.out.print((a == b) + ", ");
13      System.out.print((a != b) + ", ");
14      System.out.println(b != c);
15    }
16 }
```

程序的输出为：

true, true, true, true, false, true, true

由于关系运算符的优先级低于运算符＋，所以在第 7 行中使用括号强制先对关系运算符进行计算，然后再做字符串连接运算＋，这样就把关系运算的结果 true 或者 false 与逗号连接起来。使用 print()方法而不用 println()的目的是把所有输出放在同一行上。

2.5.4 逻辑运算符

逻辑运算符对布尔类型的变量或者字面量进行运算。逻辑运算符有：与(&&)、或(||)、非(!)。在与(&&)、或(||)、非(!)这 3 个常用的逻辑运算符中，非(!)的优先级最高，然后依次是与(&&)和或(||)。逻辑运算符及其含义如表 2-6 所示。

表 2-6 逻辑运算符

表达式	含 义
a && b	只有 a 和 b 都为 true 结果才为 true 如果 a 为 false，则直接判定结果为 false 而不对 b 计值
a || b	只要 a 和 b 中有一个为 true 结果就为 true 如果 a 为 true，则直接判定结果为 true 而不对 b 计值
! a	如果 a 为 true 则结果为 false 如果 a 为 false 则结果为 true

源文件 2-4 演示了逻辑运算符&&和||的用法。在第 5 行，由于 a==2 的值为 true，所以还需要对与运算的第二个操作数 b==3 计值，b==3 的值为 true，所以输出为 true；在第 6 行由于或运算的第一个操作数 a==2 的值为 true，第二个操作数的值无论是 true 还是 false 都对整个表达式的值没有影响了，所以就不再对第二个操作数计值，直接把第一个操作数的值作为整个表达式的值。

源文件 2-4　OperatorLogicalDemo.java

```
1 public class OperatorLogicalDemo {
2    public static void main(String[] args) {
3        int a = 2;
4        int b = 3;
5        System.out.println((a == 2) && (b == 3));
6        System.out.println((a == 2) || (b == 2));
7    }
8 }
```

程序的输出为:

```
true
true
```

2.5.5 条件运算符

条件运算符(conditional operator)是一个三目运算符(ternary operator),也就是说,该运算符有 3 个操作数。形式如下:

<条件表达式> ? <表达式 1>:<表达式 2>

首先计算<条件表达式>的值。如果为 true,则整个表达式的值是<表达式 1>的值;否则整个表达式的值是<表达式 2>的值。读作"如果<条件表达式>为真,把冒号前的表达式的值作为整个表达式的值;否则,把冒号后表达式的值作为整个表达式的值"。

例如,假设 a 的值是 9,b 的值是 8,下面的赋值语句使得 max 的值为 9:

max = (a > b) ? a : b;

条件运算符是 if-else 语句的速记形式,使得源代码可读性更好些。与赋值语句 max＝(a>b)? a:b;等价的 if-else 语句是:

```
if (a > b) {
    max = a;
}else {
    max = b;
}
```

2.5.6 位运算符

直接针对二进制位进行操作的运算符称为位运算符。位运算符有两类:按位逻辑运算符 &、|、^和 ~;移位运算符<<、>>、>>>。

一个二进制位上的数要么是 0 要么是 1。对两个二进制位 x 和 y 有进行与运算的真值表如表 2-7 所示。

对两个二进制位 x 和 y 进行或运算的真值表如表 2-8 所示。

对两个二进制位 x 和 y 进行异或运算的真值表如表 2-9 所示。

对二进制位 x 进行否运算的真值表如 2-10 所示。

表 2-7 与运算的真值表

x	y	x & y
0	0	0
0	1	0
1	0	0
1	1	1

表 2-8 或运算的真值表

x	y	x | y
0	0	0
0	1	1
1	0	1
1	1	1

表 2-9 异或运算的真值表

x	y	x^y
0	0	0
0	1	1
1	0	1
1	1	0

表 2-10 否运算的真值表

x	~x
0	1
1	0

假设有两个 N 位的二进制数 a 和 b，$a=a_1a_2\cdots a_n$，$b=b_1b_2\cdots b_n$，其按位逻辑运算的结果为 c，$c=c_1c_2\cdots c_n$。那么，如果 $a \& b = c$，则 $c_i = a_i \& b_i$；如果 $a|b=c$，则 $c_i = a_i|b_i$；如果 $\sim a = c$，那么 $c_i = \sim a_i$；其中 $1 \leq i \leq N$。

源文件 2-5 演示了按位运算。在第 3 行和第 4 行使用二进制字面量为字节类型的变量 a 和 b 赋初值。第 5 行 a & b 的计算结果为二进制 00001000，即十进制的 8；第 6 行 a|b 的计算结果为二进制 00001111，即十进制的 15；第 7 行 a^b 的计算结果为二进制 00000111，所以输出为十进制的 7；第 8 行对 a 按位非的计算结果为二进制 11110010，按补码解释，最高位为符号位，是二进制－0001110 的补码表示，所以输出十进制的－14。

源文件 2-5　OperatorBitWiseLogicalDemo.java

```
1 public class OperatorBitWiseLogicalDemo {
2     public static void main(String[] args) {
3         byte a = 0B1101;
4         byte b = 0B1010;
5         System.out.println(a  & b);
6         System.out.println(a  | b);
7         System.out.println(a  ^ b);
8         System.out.println(~a);
9     }
10 }
```

程序的输出为：

```
8
15
7
-14
```

移位运算符包括左移运算符<<、有符号右移运算符>>和无符号右移运算符>>>三种。左移运算符把二进制位向左移动指定的位数，符号位一起移动，低位补 0。

有符号右移运算符把二进制位向右移动指定的位数，符号位一起移动，高位补原符号位。

无符号右移运算符把二进制位向右移动指定的位数，高位补 0。

源文件 2-6 演示了移位运算符的使用。第 4 行中的 Integer.toBinaryString 用来按二进制字符串输出一个整数，所以程序先输出－0B00001001 的补码表示，截断高 24 位即 11110111。第 5 行左移 2 位，低位补 0，结果是 11011100，按补码解释，所以输出十进制－36；第 6 行有符号右移 2 位，当前符号位是 1，所以高位补 1，结果是 11111101，按补码解释，所以输出十进制－3；第 7 行无符号右移 00001101 两位，高位补 0，得 00000011，所以输出是十进制 3。

源文件 2-6　OperatorShiftDemo.java

```
1 public class OperatorShiftDemo {
2     public static void main(String[] args) {
3         byte a = -0B00001001;
4         System.out.println(Integer.toBinaryString(a));
5         System.out.println(a << 2);
6         System.out.println(a >> 2);
7         System.out.println(0B00001101 >>> 2);
8     }
9 }
```

程序的输出为:

```
11111111111111111111111111110111
-36
-3
3
```

2.5.7 运算符的优先级

如果在一个表达式中有多个运算符,而且没有括号显式地表明运算的顺序,那么就得应用运算符的优先级规则。比如1+2*3,先算1+2,还是先算2*3? 这取决于*的优先级是否高于+。表2-11从低优先级到高优先级的顺序列出了Java中运算符的优先次序。表示级别的数字越大,优先级越高。优先级高的运算符优先进行计算。从运算符优先级表中可以看到,*的优先级大于+,所以,在1+2*3中,应先计算2*3。

表2-11 运算符优先级

级别	类别	运算符	结合性	备注
1	赋值	=	R to L	
2	条件	?:	R to L	
3	逻辑或	||	L to R	
4	逻辑与	&&	L to R	
5	按位或	|	L to R	
6	按位异或	^	L to R	
7	按位与	&	L to R	
8	等于/不等于	== !=	L to R	
9	关系	< <= > >= instanceof	L to R	
10	移位	<< >> >>>	L to R	
11	加减	+ -	L to R	+也是字符串连接运算符
12	乘除	* / %	L to R	%整数相除取余数
13	单目	++x --x +x -x ! ~	R to L	! 逻辑非,~按位非
14	后缀	x++ x-- . []	L to R	[]访问数组元素 .访问成员

处于同一行的运算符优先级相同。括号中的运算符总是优先计算。在没有括号的情况下,如果相同优先级的运算符混合出现在表达式中,先计算哪个就取决于运算符的结合性。单目运算符和赋值运算符从右到左(R to L)计值,称为右结合;其余运算符从左到右(L to R)计值,称为左结合。例如,表达式A*B/C等价于(A*B)/C;表达式A=B=C等价于A=(B=C)。因为加减运算左结合,所以表达式3-2+1先计算3-2,再进行+运算。

括号()被用来显式地指定运算次序。例如,在如下表达式中,括号就使得先计算1+2,然后再进行 * 运算:(1+2)*3。在复杂表达式中使用括号可以提高表达式的可读性。

2.6 表达式和语句

一个字面量、一个变量就是最简单的表达式(expression),通过各类运算符连接的表达式形成新的表达式。表达式的值就是对其中字面量、变量按照运算符计值后的结果。例如,假设 x 和 y 都是 int 类型的变量,x 中已经存储了整数 2,下面都是表达式:

```
12
x
12 + x
y = x + 3
```

这些表达式的值分别是 12、2、14、5。

赋值表达式 y=x+3 还能用在另外一个表达式 z=(y=x+3)*4 中,这使得变量 y 中存放 5,变量 z 中存放 20。

表达式中还可以包含方法调用。例如表达式 5*(Math.abs(-20)/4)中包含的 Math 类中的方法 abs()调用,用来计算绝对值。

语句(statement)是以分号(;)结束的程序基本单元。比如表达式语句、变量声明语句、控制流语句等。表达式语句用来改变变量的值、调用方法和创建对象。声明语句用来声明变量。控制流语句确定语句的执行顺序。例如:

```
int i;                    //声明语句
i = 1;                    //表达式语句
if (i<10){                //控制流语句
    System.out.println(i);
}
```

使用花括号把若干条语句括起来,就成为块(block),形如:

```
{
    <语句序列>
}
```

2.7 控制台输入和输出

控制台(console)指的是与主机平台交互的命令行界面。通过 java.util.Scanner 对象可以方便地从控制台中读取基本数据类型的数据。Scanner 默认使用空白符作为分隔符把输入串分隔成词元(token)。词元即词法单元,标识符、保留字是词元;运算符是词元;逗号、分号等分隔符也是词元。空格、制表符、进纸、回车、换行等称为空白符。比如以空格作为分隔符的输入串: 12 3.14 Car,Scanner 可以从中识别为整数 12、浮点数 3.14 和字符串 Car 三个词元。源文件 2-7 演示了如何使用 Scanner 从控制台读入数据。

源文件 2-7　ScannerDemo.java

```java
import java.util.Scanner;

public class ScannerDemo {
    public static void main(String[] args) {
        Scanner sc = new Scanner(System.in);
        System.out.println("Enter an integer, a double value and
            a string:");
        int x = sc.nextInt();
        double y = sc.nextDouble();
        String s = sc.next();
        System.out.printf("You input: %d, %4.2f, %s", x, y, s);
        sc.close();
    }
}
```

System.in 是在 System 类中定义的静态变量，引用标准输入流对象，其类型为 java.io.InputStream。这个输入流默认连接到键盘，允许程序接收用户输入的数据。以其作为参数调用 Scanner 类的构造方法创建了一个 Scanner 对象用来从控制台读入数据；System.out 是 System 中定义的静态变量，引用标准输出流对象，默认连接显示器。printf() 是 out 的实例方法，提供了与 C 语言中的 printf 函数相同的语法和功能。运行程序，根据程序输出的提示"Enter an integer, a double value and a string:"输入：

```
12  3.14  Car
```

注意以空格隔开这三个数据，然后回车，程序的输出为

```
You input: 12, 3.14, Car
```

如果出现提示后输入 12，然后回车；再输入 3.14，然后回车；最后输入 Car，回车，也就是以回车作为词元的分隔符，那么 Scanner 对象同样可以读取。

Scanner 类为每一种基本数据类型都提供了相应的 nextXxx 方法用来读取该类型的值。例如 nextBoolean()、nextByte()、nextShort()、nextLong()、nextFloat()、nextDouble()、nextLine() 等。其中，nextLine 方法以字符串返回下一行。如果不是读取一整行，而是读取下一个字符串，则使用 next 方法。

还有相应的 hasNextXxx 方法，如 hasNextInt 用来判断输入串中是否还有整数。其他方法还有 boolean hasNextBoolean()、boolean hasNextByte()、boolean hasNextShort()、boolean hasNextLong()、boolean hasNextDouble()、boolean hasNextFloat()、boolean hasNextLine()、boolean hasNext()。

如果期望输入的数据类型与实际输入的数据类型不匹配，那么就会抛出 InputMismatchException 异常。例如期望输入一个整数 12，而却错误地输入了浮点数：

```
Enter an integer:3.14
Exception in thread "main" java.util.InputMismatchException
```

如果想仅读入一个字符，而不是字符串，可以使用源文件 2-8 所示的方法。

源文件 2-8　CharInputDemo.java

```java
1  import java.util.Scanner;
2
3  public class CharInputDemo {
4      public static void main(String[] args)　{
5          Scanner sc = new Scanner(System.in);
6          char ch = sc.nextLine().charAt(0);
7          sc.close();
8          System.out.println(ch);
9      }
10 }
```

其中 nextLine()方法把输入串作为字符串返回,字符串对象的 charAt(0)方法返回其中的首字符。在以 UTF-8 为默认字符编码的系统中可以直接输入和正确显示汉字。

如果想得到字符数组,那么可以使用 nextLine()方法作为字符串读入一行,然后使用字符串对象的 toCharArray()方法把字符串转换为字符数组。

2.8　控制结构

在程序运行时刻有三种不同的流程：顺序执行、选择执行和重复执行,从而形成了三种控制结构：顺序结构、分支结构和循环结构。分支结构和循环结构分别由相应的分支语句和循环语句实现。

2.8.1　顺序结构

按照语句的书写顺序逐个执行的程序执行流程形成了顺序结构。例如,假设 x 是 int 类型的变量。下面的赋值语句序列

```
x = 5;
x = x + 2;
```

的执行流程如图 2-1 所示。

图 2-1 中的图称为活动图(activity diagram)。活动图是 UML (Unified Modeling Language)的一部分,用以展示执行流程。活动图中最主要的符号就是圆角矩形框,表示一个活动。一个活动至少含有一个动作。其他常用的图形符号还有初始节点、终止节点、控制流、判定、守卫条件等。一个圆角框到另外一个圆角框的带箭头的连线表示控制流,意味着一个活动执行完毕另外一个活动才能开始。初始节点使用实心圆表示;终止节点使用有环的实心圆表示。

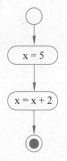

图 2-1　顺序结构

2.8.2　分支结构

分支结构也称为选择结构(selection structure)。例如,图 2-2 展示了求 x 绝对值的过程。

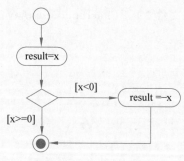

图 2-2 计算绝对值

其中的菱形表示分支节点,对应于程序中的分支语句。从分支节点出发的控制流上通常标记守卫条件。守卫条件是方括号括起来的逻辑表达式。意思是只有满足该条件,才能运行控制流通过。根据从分支节点离开的控制流是否流经语句块不同,分支结构分两种:一种由 Java 的 if 语句实现;另外一种由 if-else 语句实现。

相应的 Java 语句序列是:

```
int x = -1;
int result = x;
if (x < 0) {
    result = -x;
}
```

其中 if 语句的一般形式是:

```
if(<逻辑表达式>){
    <语句1>;
    <语句2>;
    ……
}
```

其含义是,当<逻辑表达式>为真,执行语句块;否则执行该 if 语句的下条语句。为了提高可读性,写代码时要把块中的语句使用空格缩进。

考虑求解下面的数学函数:

$$y = \begin{cases} 2x^2 + 1 & (x < 0) \\ 6x - 4 & (x \geqslant 0) \end{cases}$$

图 2-3 展示了求解过程。根据守卫条件不同,执行不同的活动。

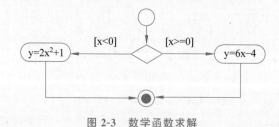

图 2-3 数学函数求解

相应的代码如下:

```
int x = -1;
int y = 0;
```

```
if (x < 0) {
    y = 2 * x * x + 1;
}else {
    y = 6 * x - 4;
}
```

代码中使用了 if-else 语句,该语句的一般形式:

```
if(<逻辑表达式>){
    <语句 1>;
    <语句 2>;
    ……
}else {
    <语句 3>;
    <语句 4>;
    ……
}
```

如果<逻辑表达式>为真,则执行前一个语句块;否则执行后一个语句块。推荐使用条件表达式而不是 if 语句以提高程序的可读性。图 2-3 的数学函数可以使用条件表达式实现求解,如源文件 2-9 所示。其中第 6 行的保留字 static 表示把方法 f()定义为类方法而不是实例方法。类方法属于类,实例方法属于对象。这样类方法 main()就能够直接调用类方法 f()而不必创建对象。

源文件 2-9　IfElseDemo.java

```
1 public class IfElseDemo {
2     public static void main(String[] args) {
3         System.out.println(f(-1));
4     }
5
6     public static int f(int x) {
7         return x < 0 ? 2 * x * x + 1 : 6 * x - 4;
8     }
9 }
```

一般地,一个独立功能由一个方法完成,该方法通过形式参数接收输入,通过返回值返回计算结果,并且方法体中没有人机交互的输入或输出。凡是需要人机交互完成输入输出的语句或者使用文件等进行输入输出的语句都在 main()方法中。

在 if-else 语句的 else 子句中还可以是 if-else 语句:

```
if (<逻辑表达式 1>) {
    ……              //块 1
} else if (<逻辑表达式 2>) {
    ……              //块 2
}
……
else if (<逻辑表达式 n>) {
    ……              //块 n
} else {
    ……              //块 n+1
}
```

在这种控制结构中,首先判断<逻辑表达式 1>是 true 还是 false,如果是 true 则执行块 1 中的语句后 if 语句结束;否则,如果是 false,判断<逻辑表达式 2>是 true 还是 false,等等,当所有的<逻辑表达式>的结果是 false 时执行块 n+1 中的语句。源文件 2-10 展示了这种控制结构。其中的类方法 getGrade 把百分制的参数 testScore 转换为等级制:ABCDF。在 main()方法中调用该方法并输出该方法的返回值。

源文件 2-10　IfElseIfDemo.java

```java
1  public class IfElseIfDemo {
2      public static void main(String[] args) {
3          System.out.println(getGrade(89));
4      }
5
6      public static char getGrade(int testScore) {
7          char grade = '\u0000';
8          if (testScore >= 90) {
9              grade = 'A';
10         } else if (testScore >= 80) {
11             grade = 'B';
12         } else if (testScore >= 70) {
13             grade = 'C';
14         } else if (testScore >= 60) {
15             grade = 'D';
16         } else {
17             grade = 'F';
18         }
19         return grade;
20     }
21 }
```

当嵌套的 if-else 语句多起来,会导致可读性下降。这种情况下建议使用另外一个能够实现分支结构的语句 switch。该语句适用于有多个分支的情况。switch 语句的一般形式:

```
switch(<表达式>){
  case <值 1>:<语句序列 1>
  case <值 2>:<语句序列 2>
    ……
  case <值 n>:<语句序列 n>
  default:<语句序列 n+1>
}
```

其中,<表达式>是整型、字符类型、枚举类型或者字符串。<语句序列>是 0 条或多条语句。当程序执行到该语句,首先对<表达式>计值,把计值结果与<值 1>进行比较,如果匹配则执行<语句序列 1>、<语句序列 2>、……、<语句序列 n>、<语句序列 n+1>;否则与<值 2>进行比较,如果匹配则执行<语句序列 2>、……、<语句序列 n>、<语句序列 n+1>;当与<值 n>匹配失败时执行 default 中<语句序列 n+1>。

源文件 2-11　SwitchDemo.java

```java
1  public class SwitchDemo {
2      public static void main(String[] args) {
3          System.out.println(toArabicNumeral('一'));
4      }
```

```
5
6    public static int toArabicNumeral(char numeral) {
7        int digit = -1;
8        switch (numeral) {
9            case '一' : digit = 1;
10           case '二' : digit = 2;
11           case '三' : digit = 3;
12           case '四' : digit = 4;
13           case '五' : digit = 5;
14           case '六' : digit = 6;
15           case '七' : digit = 7;
16           case '八' : digit = 8;
17           case '九' : digit = 9;
18           case '十' : digit = 10;
19       }
20       return digit;
21   }
22 }
```

源文件 2-11 中的类方法 toArabicNumeral() 试图把中文数字转换为阿拉伯数字。由于参数"一"与第 9 行的 case 值匹配，所以执行语句 digit=1；接着执行语句 digit=2、digit=3、……、digit=10。所以最后返回值是 10，程序的输出是 10。这并不是期望的结果。如果每种情况执行的语句互不相关，那么必须在每个 case 末尾增加 break 语句。break 语句的作用是跳出 switch 语句，去执行 switch 语句后面的语句；如果不匹配任何 <值>，则执行 "default:" 后面的语句。修改后的源代码如源文件 2-12 所示。

源文件 2-12　SwitchBreakDemo.java

```
1 public class SwitchBreakDemo {
2      public static void main(String[] args) {
3          System.out.println(toArabicNumeral('一'));
4      }
5
6      public static int toArabicNumeral(char numeral) {
7          int digit = 0;
8          switch (numeral) {
9              case '一' : digit = 1; break;
10             case '二' : digit = 2; break;
11             case '三' : digit = 3; break;
12             case '四' : digit = 4; break;
13             case '五' : digit = 5; break;
14             case '六' : digit = 6; break;
15             case '七' : digit = 7; break;
16             case '八' : digit = 8; break;
17             case '九' : digit = 9; break;
18             case '十' : digit = 10; break;
19             default: digit = -1;
20         }
21         return digit;
22     }
23 }
```

虽然可以省略 switch 语句中的 default，但是最好保留。下面的程序片段把百分制的考试分数转换成 A、B、C、D、F 和 I 等级成绩。

源文件 2-13　SwitchGradeDemo.java

```java
1 public class SwitchGradeDemo {
2     public static void main(String[] args) {
3         System.out.println(getGrade(100));
4     }
5 
6     public static char getGrade(int testScore) {
7         char grade;
8         switch (testScore / 10) {
9             case 0:
10            case 1:
11            case 2:
12            case 3:
13            case 4:
14            case 5:
15                grade = 'F';
16                break;
17            case 6:
18                grade = 'D';
19                break;
20            case 7:
21                grade = 'C';
22                break;
23            case 8:
24                grade = 'B';
25                break;
26            case 9:
27            case 10:
28                grade = 'A';
29                break;
30            default:
31                grade = 'I';
32        }
33        return grade;
34    }
35 }
```

在源文件 2-13 中,首先对表达式 testScore / 10 计值,其结果是一个整数。然后从第一个 case 开始与其后的整数进行匹配。最后执行匹配成功的 case 后面的语句序列。

因此,如果 testScore 的值是 100,那么执行 case 10 后面的语句序列:

```
grade = 'A';
break;
```

使得 grade 的值是 'A'。

如果 testScore 的值是 90,那么 grade 将被赋值 'A'。

源文件 2-13 中的 switch 语句有两个问题:一是从第 9 行到第 14 行共享第 15 和第 16 行的语句,但读起来令人觉得莫名其妙;二是太多的 break 语句,读起来让人觉得啰嗦。可以使用"标签规则"来解决这两个问题。源文件 2-14 展示了应用"标签规则"的百分制成绩转换为等级制,比源文件 2-13 的可读性提高了很多。

源文件 2-14 SwitchGradeArrowLabelDemo.java

```java
1  public class SwitchGradeArrowLabelDemo {
2      public static void main(String[] args) {
3          System.out.println(getGrade(100));
4      }
5
6      public static char getGrade(int testScore) {
7          char grade;
8          switch (testScore / 10) {
9              case 0, 1, 2, 3, 4, 5 -> grade = 'F';
10             case 6                -> grade = 'D';
11             case 7                -> grade = 'C';
12             case 8                -> grade = 'B';
13             case 9, 10            -> grade = 'A';
14             default               -> grade = 'I';
15         }
16         return grade;
17     }
18 }
```

当运行时一旦<表达式>与箭头->左侧的任一标签匹配,将运行位于箭头右侧语句,并且不再运行任何其他语句。一个箭头->连接的左部和右部称为一条 switch 标签规则。箭头左部表示规则的条件;箭头右部则是规则的动作:

```
switch (<表达式> ) {
    case <左部 1> -> <右部 1>
    case <左部 2> -> <右部 2>
        ……
    case  <左部 n> -> <右部 n>
    default <左部 n+1> -> <右部 n+1>
}
```

其中,<左部>是逗号隔开的若干字面量;<右部>是一条语句或一个语句块。

源文件 2-15 首先输出了一个具有 4 个选项的菜单,在第 21 行使用标签规则根据用户输入的菜单项索引号选择不同的功能去执行。

源文件 2-15 Menu.java

```java
1  import java.util.Scanner;
2
3  public class Menu {
4      public static void main(String[] args) {
5          //菜单项编号,用户可以使用菜单项编号选择菜单
6          int optionNumber;
7          String message = "Nothing to do";
8          Scanner sc = new Scanner(System.in);
9          /* 显示菜单,得到用户输入 */
10         System.out.println("""
11             1. insert
12             2. delete
13             3. update
14             4. query
15
```

```
16          Enter the number of your choice:
17          """);
18
19       optionNumber = sc.nextInt();
20
21       switch (optionNumber) {
22          case 1 -> message = "insert something";
23          case 2 -> message = "delete something";
24          case 3 -> message = "update something";
25          case 4 -> message = "query something";
26          default -> {
27             System.out.println("Illegal option. Quit!");
28             System.exit(1);
29          }
30       }
31       System.out.println(message);
32    }
33 }
```

标签规则的左部可以是若干个值。如有允许用户通过输入索引号,或者大写的首字母,或者小写的选项来确定选项,可使用具有三个值的左部,如源文件 2-16 的第 21 行所示。

源文件 2-16　Menu2.java

```
1 import java.util.Scanner;
2
3 public class Menu2 {
4    public static void main(String[] args) {
5       //菜单项编号,用户可以使用菜单项编号选择菜单
6       String option;
7       String message = "Nothing to do";
8       Scanner sc = new Scanner(System.in);
9       /* 显示菜单,得到用户输入 */
10      System.out.println("""
11             1. Insert
12             2. Delete
13             3. Update
14             4. Query
15
16          Enter the number of your choice:
17          """);
18
19       option = sc.next();
20
21       switch (option) {
22          case"1", "I", "insert" -> message = "insert something";
23          case"2", "D", "delete"  -> message = "delete something";
24          case"3", "U", "update"  -> message = "update something";
25          case"4", "Q", "query"   -> message = "query something";
26          default -> {
27             System.out.println("Illegal option. Quit!");
28             System.exit(1);
29          }
30       }
```

```
31          System.out.println(message);
32      }
33 }
```

2.8.3 循环结构

循环结构用于重复执行语句块。Java 中有三种语句实现循环结构：while 循环、do-while 循环和 for 循环。while 语句的一般形式为：

```
while(<循环条件>){
    <语句块>
}
```

意思是当<循环条件>为真时执行<语句块>。执行步骤如下：

（1）对<循环条件>计值；

（2）如果<循环条件>为假，跳过<语句块>，执行随后的语句；

（3）如果<循环条件>为真，执行<语句块>；

（4）转到步骤(1)。

图 2-4 展示了把公差 d 为 1、首项 a 为 1、项数 n 为 10 的等差数列累加的过程。源文件 2-17 展示了如何应用 while 循环实现累加。

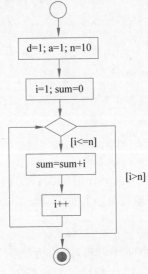

图 2-4 累加

源文件 2-17 **WhileLoopDemo.java**

```
1 public class WhileLoopDemo {
2     public static void main(String[] args)   {
3         int a = 1;            //首项
4         int n = 10;           //等差数列中的前 n 项
5         int d = 1;            //公差
6         int sum = 0;
7         int i = 1;
8         while (i <= n) {
```

```
9          sum = sum + (a + (i-1) * d);
10         i++;
11      }
12   }
13 }
```

一般来说，在循环结构中有 4 个部件：初始化、循环条件、循环体和更新循环条件，如图 2-5 所示。"初始化"部分完成反复执行语句块前的准备工作；保留字 while 后面的逻辑表达式称为"循环条件"；将被反复执行的语句块称为"循环体"。

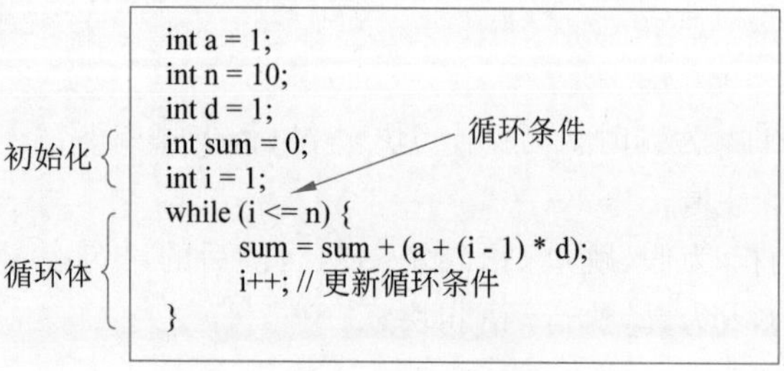

```
             int a = 1;
             int n = 10;
             int d = 1;
初始化 {      int sum = 0;
             int i = 1;          循环条件
             while (i <= n) {
循环体 {         sum = sum + (a + (i - 1) * d);
                 i++; // 更新循环条件
             }
```

图 2-5 循环控制结构的组成

由于"循环条件"控制了循环体的执行，所以在循环体中必须有更新循环条件的语句。否则，如果<循环条件>总是计值为真，那么循环体将被无限执行。这种情况俗称"死循环"。下面的程序片段演示了"死循环"：

```
int a = 1;
int n = 10;
int d = 1;
int sum = 0;
int i = 1;
while (i <= n) {
    sum = sum + (a + (i-1) * d);
}
```

由于在循环体中没有更新变量 i，所以 i 的值保持初值 1 不变；从而使得每次对循环条件"i<=n"计值为真，从而造成不停地反复执行循环体：

```
sum = sum +(a + (i-1) * d);
```

循环体中还可以包含另外一个循环语句，这种情况称为嵌套循环(nested loop)。在嵌套循环中，如果外循环需要执行 M 次，内循环需要执行 N 次，那么外循环的每次执行，内循环都要执行 N 次。即内循环总共执行了 M×N 次。

do-while 语句的语法：

```
do{
    <语句块>
}while(<循环条件>);
```

do-while 反复执行循环体，直到<循环条件>计值为 false。因此，do-while 循环中的循环体至少被执行 1 次。源文件 2-18 演示了如何使用 do-while 循环对首项为 1 公差为 1 的 n 项等差数列累加。

源文件 2-18　DoWhileLoopDemo.java

```
1  public class DoWhileLoopDemo {
2      public static void main(String[] args)  {
3          int a = 1;           //首项
4          int n = 10;          //等差数列中的前 n 项
5          int d = 1;           //公差
6          int sum = 0;
7          int i = 1;
8          do {
9              sum = sum +(a + (i - 1) * d);
10             i++;
11         } while (i <= n);
12     }
13 }
```

如果 n 的初值为 0 而不是 10，那么源文件 2-17 中 while 循环由于循环条件为假，无法执行循环体，累加和 sum 的结果就是初值 0，结果正确；而源文件 2-18 使用了 do-while 循环，先累加了首项，再判断循环条件，虽然循环条件为假，但是已经累加了首项，出现了错误的结果。所以，要在确认了至少执行一次循环体的情况下再使用 do-while 循环。

while 循环和 do-while 循环适合于无法预知循环次数的情况。如果预先知道循环次数，则使用 for 循环使得代码更为简洁些。for 语句的语法：

```
for(<初始化>;<循环条件>;<更新循环条件>){
    <语句块>
}
```

先执行<初始化>，如果<循环条件>为真则执行循环体、执行<更新循环条件>；否则 for 语句结束，执行后续语句。

源文件 2-19 演示了如何使用 for 循环对首项为 1，公差为 1 的 n 项等差数列累加。

源文件 2-19　ForLoopDemo.java

```
1  public class ForLoopDemo {
2      public static void main(String[] args)  {
3          int a = 1;           //首项
4          int n = 10;          //等差数列中的前 n 项
5          int d = 1;           //公差
6          int sum = 0;
7          for (int i = 1; i <= n; i++) {
8              sum = sum +(a + (i - 1) * d);
9          };
10     }
11 }
```

for 循环是一种"all in one"的循环控制结构：把组成循环四个部件中的三个组合到了 for 后的圆括号中。在源文件 2-19 中，for 循环的执行次序如下：

(1) 执行初始化语句 i=1，把 1 赋值给整型变量 i。

(2) 对循环条件 i<=n 计值。如果为真，执行循环体；否则结束 for 语句执行随后的语句。

(3) 执行循环体中的语句：sum＝sum＋(a＋(i－1)＊d)。

(4) 执行<更新循环条件>：i＋＋。

(5) 转步骤(2)。

这三个循环语句在功能上是等价的。如果事先知道至少执行一次循环体,那么do-while语句是比较好的选择;如果事先知道循环体执行的次数,那么通常选择使用for语句。

在if/else/for/while/do语句中必须使用花括号,即使只有一条语句。这不是语法要求,而是可读性要求。

2.8.4 分支语句

Java中有三个专门用于控制程序执行流程的语句,称为分支语句(branching statement):break、continue和return。

在前面介绍的switch语句中,break语句使得程序的执行流程立即从该语句位置跳转到switch语句的后随语句,即终止了switch语句的执行转去执行其后随语句。break也可放在for语句、while语句或者do-while语句的循环体中用以终止循环,转去执行循环语句的后随语句。例如下面的代码片段

```java
int i = 9;
while (i >= 1) {
    i--;
    if (i == 3) {
        break;
    }
    System.out.printf("%d ", i);
}
```

的输出结果是87654,而不是87654321。因为每次执行循环体都会判断i==3,如果计值为真,则执行break跳出循环体结束程序执行。

continue语句仅跳过本次执行循环体时尚未执行的语句,进行下次执行循环体。例如下面的代码片段

```java
int i = 9;
while (i >= 1) {
    i--;
    if (i == 3) {
        continue;
    }
    System.out.printf("%d ", i);
}
```

的输出结果是87654210,而不是876543210。因为每次执行循环体都会判断i==3,如果计值为真,则执行continue跳过本次执行循环体剩余语句"System.out.printf("%d ", i);",继续进行下次循环体的执行。

return语句用以从当前正在执行的方法中返回到调用该方法的位置。除了能够将控制流返还给方法的调用者,return语句还能交给方法的调用者一个值,称为返回值。仅需要把要返回的值放在关键字return的后面即可。例如语句return ++i;返回对变量i增1后的结果。语句return "Donald";返回字符串"Donald"的引用。返回值的数据类型必须和方法声明的返回值类型兼容。如果方法声明的返回值类型是void,那么就不能返回任何值。

2.9 数组

2.9.1 数组的概念

数组是线性表的顺序存储结构实现,使用连续的存储单元存放具有相同数据类型的多个数据。一个数组就是一个对象。使用数组有以下 4 个步骤。

(1) 声明数组类型的引用变量;
(2) 创建数组对象,并通过前面声明的变量进行引用;
(3) 初始化数组对象;
(4) 访问数组对象。

2.9.2 数组的声明

声明数组的语法一般是:

```
<元素数据类型>[]   <变量名>
```

例如:

```
int[] a;
```

声明了元素数据类型为 int 的数组引用,在名字为 a 的变量中存储该数组的引用。注意此处的"a"并不是实际存储数据元素的存储空间标识,而是该存储空间的引用。存储引用的变量称为引用类型变量。

int[] a 仅仅声明了数组引用变量,并没有为数组分配存储空间,还需要使用保留字 new 创建实际存储数组数据元素的存储空间。例如:

```
int[] a;              //声明引用数组对象的变量
a = new int[6];       //创建数组对象
```

"new int[6]"的意思是创建容量为 6 的数据元素类型为 int 的数组对象。

"a=new int[6]"中赋值的作用是把创建容量为 6 的数据元素类型为 int 的数组对象的引用赋值给变量 a。

可以把上面的两条语句合体:

```
int[] a = new int[6];    //声明并创建数组对象
```

图 2-6 展示了"int[] a=new int[6]"执行后存储空间的布局。从中可以看到,数组对象中不仅有元素的顺序存储空间,还有一个属性 length,其值是数组的容量。

如果没有对数组进行显式初始化,那么 Java 会把每个数组元素设置为其数据类型的默认值。整数类型的默认值是 0,浮点数类型的默认值是 0.0,boolean 类型的默认值是 false,char 类型的默认值是 '\u0000'。

在声明数组时可以使用如下语法设置初值:

```
数据类型[] 变量名 = {元素 1,元素 2,...,元素 n};
```

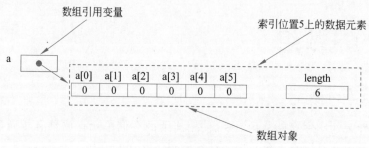

图 2-6　数组对象及其引用

例如：

```
int[] a = {92, 83, 74};
```

这条语句首先以花括号中列举的字面量作为初值，以其个数作为数组的容量创建了一个数组对象，然后把这个数组对象的引用赋值给变量 a。该数组对象的内存布局如图 2-7 所示。从中可以看到，按照字面量的顺序，数组元素依次安排在连续的存储空间，并且每个元素都有一个索引，索引号从 0 开始。

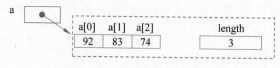

图 2-7　数组声明时设置初值

也可以在创建数组对象时通过列举字面量作为数组对象的初值：

```
int[] a;
a = new int[]{1, 3 , 5, 6};
```

由于数组的容量是个 int 类型的整数，所以，其最大值就被限制为 int 类型的最大值：2147483647。

2.9.3　数组的访问

创建数组对象并建立对数组对象的引用后，就可以使用该数组对象了。通过数组中每个元素的索引（index），或者称为下标（subscript）来访问数组元素。数组对象的成员变量 length 记录着数组的容量，通过成员访问运算符"."能够获取这个数：

```
int arraySize = a.length;
```

一个数组对象的有效索引值从 0 开始到数组的长度 length－1。例如下面的代码片段首先声明和创建了一个具有 4 个元素的整型数组（容量为 4），其有效索引分别是 0、1、2 和 3。通过索引就可以把数组中的第一个元素（索引是 0）修改为 10，把最后一个元素（索引是 3）输出：

```
int[] a = {1, 3, 5, 6};       //具有 4 个元素的数组
a[0] = 10;                    //把 10 赋值给索引为 0 的元素
System.out.print(a[3]);       //输出索引为 3 的数组元素
```

使用无效的索引访问数组会产生索引越界异常。例如，源文件 2-20 第 4 行试图访问 a[4] 而最大索引号是 3，4 是无效索引，所以程序抛出运行时刻异常，控制台会显示：

```
Exception in thread "main" java.lang.ArrayIndexOutOfBoundsException: Index 4
out of bounds for length 4 at ArrayExceptionDemo.main(ArrayExceptionDemo.java:
4)
```

意思是，在 ArrayExceptionDemo.java 的第 4 行 main 方法中产生了数组索引越界异常：索引 4 超出了 4 个元素数组的索引范围。

源文件 2-20　ArrayExceptionDemo.java

```
1 public class ArrayExceptionDemo {
2     public static void main(String[] args)  {
3         int[] a = {1, 3, 5, 6};
4         a[4] = 10;
5     }
6 }
```

数组元素的数据类型也可以是引用类型。在这种情况下，被引用的对象必须被事先创建。也就是说，数组中存放的仅仅是对象的引用，而不是对象本身。

例如，Java 把字符串按照对象处理，下面的程序片段首先根据花括号中列举的 4 个字符串（String）字面量创建了 4 个字符串（String）对象，然后将这 4 个字符串对象的引用依次存放在数组中。

```
String[] colors = {"red", "orange", "yellow", "green"};
for(String s: colors){
    System.out.println(s);
}
```

图 2-8 展示了这个字符串数组的内存布局。

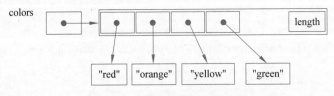

图 2-8　字符串数组的内存布局

上面程序片段中的 for 循环是专门用来遍历集体（collection）的循环语句，称为 for-each 语句。若干个数据项或者对象就组成一个集体。for-each 循环用以遍历集体。其语法为：

```
for (<类型><元素> : <集体>){
    <语句块>
}
```

其中，<类型>指数组元素的类型，<变量>用以存储当前访问的集体中的对象。读作：对于<集体>中每个<元素>执行循环体。所以

```
for(String s: colors){
    System.out.println(s);
}
```

就应读作：对于数组 colors 中的每个字符串引用 s 执行循环体 System.out.println(s);

按照英语读法，for-each 语句中的冒号":"应读作 in。整个语句应读作：

for each String s IN colors, do loop body System.out.println(s);

2.9.4 二维数组

如果数组的元素是对另外数组的引用，那么就构成了多维数组（multidimensional array）。假设对如下矩阵中的全部元素求和：

$$\begin{matrix} 1 & 2 & 3 \\ 4 & 5 & 6 \end{matrix}$$

这个矩阵有 2 行 3 列。可以把每一行看作一个一维数组，这样就有了两个一维数组；然后把这两个一维数组对象再存储到另一个一维数组中，如图 2-9 所示。

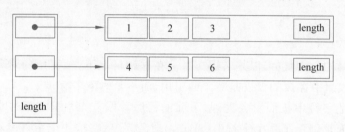

图 2-9 二维数组的内存布局

源文件 2-21 演示了使用二维数组作为矩阵的存储结构来累加矩阵中的所有元素。

源文件 2-21 Array2DimDemo.java

```
1  public class Array2DimDemo {
2      public static void main(String[] args)   {
3          int[][] matrix = {{1, 2, 3}, {4, 5, 6}};
4          int sum = 0;
5
6          //使用 for-each 循环遍历数组元素
7          for (int[] row : matrix) {      //对于二维数组中的每一行
8              for (int e : row) {          //对于一行中的每个元素
9                  sum += e;
10             }
11         }
12         System.out.println("Summation: " + sum);
13     }
14 }
```

在第 3 行声明的 matrix 本质上引用一个一维数组，其每个元素还是数组对象：{1, 2, 3}和{4, 5, 6}。

内嵌的数组对象不必等长。源文件 2-22 中的代码展示了如何累加不等长的二维数组中的所有元素。

源文件 2-22 Array2DimUnequalLength.java

```
1  public class Array2DimUnequalLength {
2      public static void main(String[] args) {
3          int[][] a = {{1, 2, 3, 4}, {5, 6, 7}, {8, 9}};
```

```
4        int sum = 0;
5        for (int[] row: a){
6            for(int e: row){
7                sum += e;
8            }
9        }
10       System.out.println(sum);
11   }
12 }
```

在源文件 2-22 的代码中，首先通过在声明数组类型的引用变量时列举字面量的办法创建了二维数组对象并通过变量 a 来引用。这个二维数组对象有 3 个元素，每个元素都是一维数组，内存布局如图 2-10 所示。

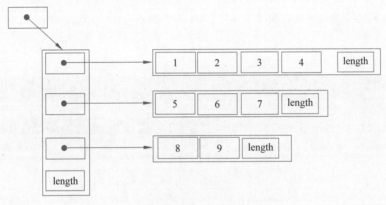

图 2-10 以不等长数组为元素的二维数组

然后程序使用 for-each 循环遍历该二维数组：外循环通过 int[] 类型的引用变量 row 引用各个不等长的一维数组；内循环通过 int 类型的引用变量 e 引用一维数组 row 中的各个整数。

Java API 中的 java.util.Arrays 类包含了对数组操作的各种各样的静态方法。例如使用静态方法 sort() 对数组排序：

源文件 2-23　ArraysSortDemo.java

```
1 public class ArraysSortDemo {
2    public static void main(String[] args)  {
3        int[] numbers = {6, 4, 1, 2, 2, 3, 5};
4        java.util.Arrays.sort(numbers);
5        for (int number : numbers) {
6            System.out.printf("%2d", number);
7        }
8    }
9 }
```

当调用其他类中的静态方法时要在方法前增加类名，如源文件 2-23 第 4 行所示。如果在源文件的第 1 行使用保留字 import 导入了 java.util.Arrays 类，那么每次调用该类中的静态方法时就可以省略类名前的包名 java.util，如源文件 2-24 第 1 行和第 6 行所示。

源文件 2-24. ArraysSortImportDemo.java

```
1 import java.util.Arrays;
2
3 public class ArraysSortImportDemo {
4     public static void main(String[] args)   {
5         int[] numbers = {6, 4, 1, 2, 2, 3, 5};
6         Arrays.sort(numbers);
7         for (int number : numbers) {
8             System.out.printf("%2d", number);
9         }
10    }
11 }
```

从官方的帮助文档主页（https://docs.oracle.com/en/java/javase/17/docs/api/index.html）查看其他的静态方法的名字、功能描述以及具体用法，如图 2-11 所示，在 SEARCH 文本输入框中输入类名 arrays，就会出现候选项，从中选择 java.util.Arrays 就会打开 Arrays 类的帮助文档。

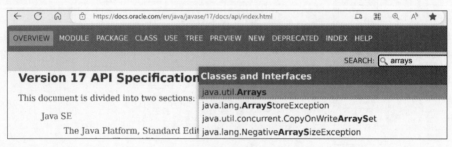

图 2-11　API documentation

使用静态方法 Arrays.equals()来比较两个数组是否相等。如果两个数组的长度 length 相同，而且对应元素相等，那么这两个数组相等。下面语句创建的数组中，数组 a 和 b 相等，而 b 和 c 不相等：

```
int[] a = {2, 4, 5, 6};
int[] b = {2, 4, 5, 6};
int[] c = {3, 4, 5, 6};
System.out.println(java.util.Arrays.equals(a, b));      //真
System.out.println(java.util.Arrays.equals(b, c));      //假
```

下面语句创建的两个数组也相等，程序会输出 true。

```
String[] dinnerA = {"Soup", "Mushroom", "Seasonal Fruit"};
String[] dinnerB = {"Soup", "Mushroom", "Seasonal Fruit"};
System.out.println(java.util.Arrays.equals(dinnerA, dinnerB));
```

第 2 章　章节测验

第3章 类和对象

Java 程序通过对象及对象间的协作完成一项功能。为了让对象能做些事情,就得先创建对象。而创建对象的依据就是"类",类相当于对象的模板,描述了对象由哪些成分组成,能够提供哪些功能。Java 程序就是一些类的集合。设计 Java 程序就是设计和使用类。通常一个类保存到一个.java 源文件中,如 A.java。所谓设计"类",就是把所关注的现实世界中事物的特征及关系抽取出来,形成类模型,然后使用 Java 语言实现该类模型。

3.1 类的声明

从静态角度看,面向对象的程序是由类组成的;从动态角度(程序运行时刻)看,面向对象程序依靠对象及对象间的方法调用(或称消息传递)来完成需要的功能。

类是对象的抽象表示,相当于一个概念;而对象是类的实例,相当于一个具体的事物。例如,"汽车"是一个"类",包含了品牌、型号、排量、加速、刹车等特征,某一辆具体的小汽车就是"汽车"这个类的一个实例。对象具有唯一的标识(identity)、状态(state)和行为(behavior)。对象的状态由其成员变量(member variable)以及这些成员变量当前的值表示。成员变量定义了对象的属性。对象的行为由一系列方法(method)来定义。调用对象上的方法意味着请求该对象完成一项工作。

图 3-1 使用 UML 类图展示了小汽车 Car 的特征,包括车牌号码 licenceNumber、车身外廓尺寸(长 length、宽 width、高 height)和当前行驶速度 speed 等属性;另外还有加速 accelerate、获取外廓尺寸 getOverallSize 和鸣笛 honk 三个行为。其中-表示类私有,只允许在类内访问;+表示公共,允许任意类访问。类的属性和行为描述在 Java 类声明中。

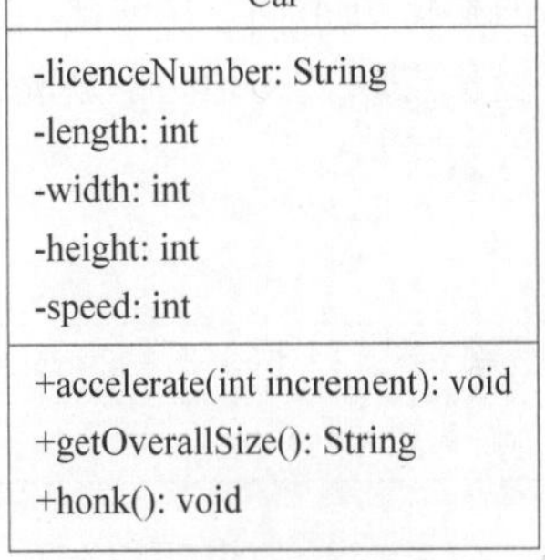

Car
-licenceNumber: String
-length: int
-width: int
-height: int
-speed: int
+accelerate(int increment): void
+getOverallSize(): String
+honk(): void

图 3-1 Car 的类模型

在类的声明中，包括保留字 class、类名和类体等部分，类体中包含了成员变量、成员方法和构造方法等成员，如图 3-2 所示。构造方法（constructor）是 Java 类声明中的特殊方法：仅在创建对象时被调用，用于初始化对象的状态。构造方法的名字与类名完全相同，另外，构造方法无返回值类型声明。

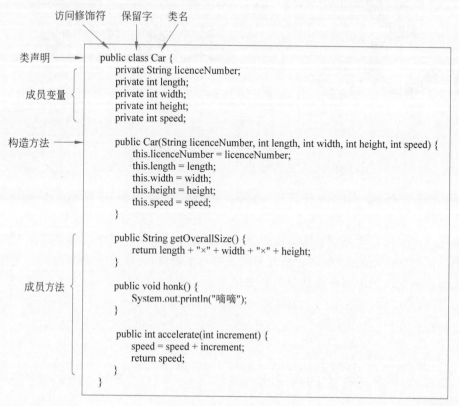

图 3-2 类的组成

类的属性声明为成员变量，变量的类型可以是基本数据类型，也可以是引用类型。

类的行为声明为成员方法。成员方法由方法头和方法体组成，能够被调用，能够接收参数，能够返回给调用者一个值。

如果在声明成员变量的时候程序员没有显式地给出成员变量的初值，那么使用默认的初值。例如在图 3-2 中，成员变量 speed 没有初值，则使用默认的初值 0。数据类型不同，默认的初值也不同。8 种基本数据类型以及引用类型的默认初值如表 3-1 所示。注意，仅成员变量有默认初值，而局部变量（在方法体中声明的变量）没有默认初值，其初值必须被显式指定。

表 3-1 默认初值

数据类型	byte	short	int	long	float	double	char	boolean	引用
默认初值	0	0	0	0L	0.0f	0.0d	'\u0000'	false	null

成员变量的声明语句中还有对变量可访问性的声明。对变量的读写合称为对变量的访问。在图 3-2 中，成员变量前的访问修饰符 private 的意思是仅仅在本类中可视，称为类私

有变量。访问修饰符共有 3 个：public、protected 和 private。如果没有访问修饰符，则使用默认的"包私有"可访问性。

成员方法也有不同的可访问性，如果声明为 public，则表示该方法允许被任意类调用。对方法的调用，也称为对方法的访问。

声明方法的格式如下：

```
<访问修饰符><返回值类型><名字>(<形式参数序列>) {
    <语句序列>
}
```

其中，<形式参数序列>是若干个逗号隔开的<类型><形式参数名>。<返回值类型>、<名字>和(<形式参数序列>)合起来称为方法头部(method head)。方法头部中，<名字>和<形式参数类型序列>合起来称为方法签名(method signature)，唯一标识了同一类中的方法。<形式参数类型序列>是逗号隔开的若干形式参数类型，不含形式参数名。例如构造方法 Car(String licenceNumber, int length, int width, int height, int speed)的形式参数类型序列是(String, int, int, int, int)。形式参数的名字、<访问修饰符>以及<返回值类型>不在方法签名中。形式参数的个数不同、对应参数类型不同，那么方法签名也不同。使用方法签名作为方法的标识，使得同一个类中可以声明多个相同名字的方法。

方法声明中花括号括起来的部分称为方法体。方法体中声明的变量称为局部变量(local variables)。这些变量仅能被局部于方法体的语句使用。而且必须被显式地赋初值。在方法体中通过 return <值>语句返回给方法的调用者一个值。该值的数据类型在方法头部中的<返回值类型>部分声明。如果一个方法不返回任何值，则把这个方法的返回值类型声明为 void。图 3-3 展示了方法声明中的各个部分。

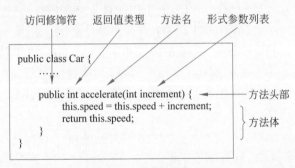

图 3-3　方法声明

如果在一个类中没有声明构造方法，则是使用默认的无参构造方法；一旦在类中声明了构造方法，无论该构造方法有无参数，该类都不再使用默认的无参构造方法。图 3-2 类 Car 中声明了一个有参构造方法 Car (String licenceNumber, int length, int width, int height, int speed)，而在源文件 3-1 中的 CarDemo 类中却调用了无参构造方法 Car()，因为无参构造方法不存在，就会出现编译错误：

```
CarDemo.java:3: error: constructor Car in class Car cannot be applied to given
types;
        Car a = new Car();
                ^
  required: String, int, int,int, int
```

```
found: no arguments
reason: actual and formal argument lists differ in length
1 error
```

意思是在源文件 CarDemo.java 第 3 行出现错误：类 Car 中的构造方法 Car 实际参数（actual argument）与形式参数（formal argument）个数不同。

源文件 3-1　CarDemo.java

```
1 public class CarDemo {
2     public static void main(String[] args) {
3         Car a = new Car();
4     }
5 }
```

除了通过保留字 class 自定义类，JDK 中还有大量已经定义好的类用来完成通用的功能，例如用于字符串处理的 String 类，用来从字符流中读入数据的 Scanner 类等。此外，还可以使用第三方的类库。

3.2　创建对象

声明了类就为创建对象准备好了模板。模板是对一个事物的抽象描述；而对象是具体的一个事物。有了类，就可以根据类创建对象了。创建对象首先要在计算机的存储空间中为对象分配一块区域。

在 Java 程序运行时刻有两块内存区域：一个是栈（Stack），另一个是堆（Heap）。在栈中存放方法中定义的基本类型变量和对象的引用变量；在堆里存放对象。

例如，源文件 3-2 中的 Car 类声明了三个成员变量：长 length、宽 width 和高 height；声明了两个成员方法：getOverallSize 返回外廓尺寸字符串，honk 以字符串输出模拟的汽车鸣笛。在第 3 行创建了一个具体的 Car 类型的对象，并将这个对象的引用赋值给引用类型变量 a，如图 3-4 所示，栈区中的变量 a 中存放了堆区中对象的引用。3 个整型成员变量的存储空间形成了对象的存储空间。

源文件 3-2　ObjectCreation.java

```
1 public class ObjectCreation {
2     public static void main(String[] args) {
3         Car a = new Car(4481, 1746, 1526);
4     }
5 }
6
7 class Car {
8     private int length;
9     private int width;
10    private int height;
11
12    Car(int length, int width, int height) {
13        this.length = length;
14        this.width = width;
15        this.height = height;
16    }
```

```
17
18      String getOverallSize() {
19          return length + "×" + width + "×" + height;
20      }
21
22      void honk() {
23          System.out.println("嘀嘀");
24      }
25
26 }
```

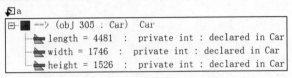

图 3-4 Car 对象及其引用

第 3 行中 Car 是在同一个源文件中定义的类,Car(4481,1746,1526)是对构造方法 Car(int length,int width,int height)的调用,保留字 new 的意思是根据 Car 类型声明分配堆存储空间,并调用构造方法 Car(4481,1746,1526)初始化分配的存储空间。

new 返回所创建对象的引用(reference)。引用是对象的标识,以便该对象被访问,从实现角度看,为对象所分配存储区域的起始地址即可作为引用。

因此,创建一个对象需要做三件事情:声明引用变量,实例化(instantiation)对象,初始化(initialization)对象。这三件事情可以使用一条语句完成,如下所示:

$$\underbrace{Car \quad a}_{\text{声明引用变量}} = \underbrace{new}_{\text{实例化}} \underbrace{Car(4481, 1746, 1526);}_{\text{初始化}}$$

声明引用变量意味着在栈空间分配一个变量,并使用一个名字标识该变量;该变量中允许存放指定类型的对象的堆空间引用。实例化的含义是使用运算符 new 为对象分配堆存储空间。初始化的含义是调用构造方法执行相应的初始化操作。创建对象的语句:

```
a = new Car(4481, 1746, 1526);
```

读作:创建类型为 Car 的对象,并将其引用赋值给变量 a。
英文读作:create an object of type Car and assign its reference to a。
也可以分成两条语句完成对象创建:

```
Car a;
a = new Car(4481, 1746, 1526);
```

构造方法 Car(int length,int width,int height)中的形式参数与成员变量一一对应,称为"全参数构造方法"。

3.3 访问对象

创建了对象后,通过其引用就可以访问对象了。例如在源文件 3-3 第 3 行创建了对象并通过 a 来引用该对象,第 4 行让 a 所引用的对象执行方法 honk()输出字符串"嘀嘀";第 5

行让 a 所引用的对象执行方法 getOverallSize() 返回并输出汽车的外廓尺寸"4481×1746×1526"。

源文件 3-3　ObjectVisit.java

```java
1 public class ObjectVisit {
2     public static void main(String[] args) {
3         Car a = new Car(4481, 1746, 1526);
4         a.honk();
5         System.out.println(a.getOverallSize());
6     }
7 }
8
9 class Car {
10     private int length;
11     private int width;
12     private int height;
13
14     Car(int length, int width, int height) {
15         this.length = length;
16         this.width = width;
17         this.height = height;
18     }
19
20     String getOverallSize() {
21         return length + "×" + width + "×" + height;
22     }
23
24     void honk() {
25         System.out.println("嘀嘀");
26     }
27
28 }
```

成员访问运算符"."用来通过对象引用调用对象的方法。如果该方法没有声明形式参数，则调用格式形如：

<对象引用>.<方法名>()

如果该方法声明了形式参数，则调用时需要给出相应的实际参数（argument）序列：

<对象引用>.<方法名>(<参数序列>)

同样通过操作符"."，对象引用可以访问对象的成员变量。但是，通常成员变量声明为类私有的，只能通过成员方法间接访问。

读写对象的成员变量、执行对象的方法统称为访问对象的成员。访问对象的成员使用成员访问运算符"."。该运算符前是对象的引用。所以，在前面的例子中先创建对象，得到对象的引用后再通过对象引用访问对象。那么，在对象的构造方法中和对象的成员方法中如何访问该对象的成员变量呢？

在源文件 3-3 的第 14 行到第 18 行声明了一个构造方法，该方法把长、宽、高三个实际参数赋值给成员变量。所以在方法体中就需要访问成员变量，但是，此时尚没有对象引用，如何"通过对象引用来访问成员"呢？保留字 this 就来解决这个问题。构造方法中的 this

引用正是由保留字 new 创建的对象。第 3 行中的保留字 new 后面是对第 14 行构造方法的调用，调用时传递了三个参数，然后控制转移到第 15 行开始执行，第 15 行 this.length＝length 把实际参数 length 赋值给正在创建的对象的成员变量 length。

在对象的方法中访问成员变量仍然需要通过对象引用，仍然使用保留字 this。例如在第 20 行的 getOverallSize()方法中需要读取三个成员变量，应该写作：

```
public String getOverallSize() {
    return this.length + "×" + this.width + "×" + this.height;
}
```

此时 this 引用的是在第 5 行 a.getOverallSize()中的 a 所引用的对象。即 this 总是引用执行 this 所在方法的那个对象。通常把方法体中的 this 省略以提高可读性。

通过 this()就可以在构造方法中显式地调用一个构造方法。在源文件 3-4 中有两个构造方法，其中 Car(String licenceNumber)显式地调用了构造方法 Car(String licenceNumber,int length,int width,int height)：

```
this(licenceNumber, 0, 0, 0);
```

意思是调用当前对象所属类的方法签名为 Car(String，int，int，int)的构造方法。注意，在构造方法的方法体中，this()必须是第一条语句。

源文件 3-4　Car.java

```
public class Car {
    private String licenceNumber;
    private int length;
    private int width;
    private int height;

    public Car(String licenceNumber, int length, int width, int height) {
        this.licenceNumber = licenceNumber;
        this.length = length;
        this.width = width;
        this.height = height;
    }

    public Car(String licenceNumber) {
        this(licenceNumber, 0, 0, 0);
    }
}
```

3.4　对象的字符串表示

在源文件 3-5 中第 3 行、第 4 行和第 5 行分别声明了一个初值为 45 的整型变量 a、引用字符串对象"ABC"的引用变量 s 和引用 Car 对象的引用变量 c。然后使用 System.out.println()输出这三个变量，结果如下：

```
45
ABC
Car@28a418fc
```

println()方法输出了变量a的值45、变量s引用的字符串对象"ABC"。但是，变量c所引用的Car对象却输出为"Car@28a418fc"，意思是在内存地址28a418fc处的Car对象。这往往不是期望的输出。假如期望的输出形如：

```
Car [length: 4481, width: 1746, height: 1746]
```

那么就需要在Car类中增加一个头部为public String toString()的方法，如源文件3-6中第31行所示。该方法由println()方法隐式调用，即语句：

```
System.out.println(c);
```

等价于语句：

```
System.out.println(c.toString());
```

源文件3-5　ObjectByString.java

```
1  public class ObjectByString {
2      public static void main(String[] args) {
3          int a = 45;
4          String s = "ABC";
5          Car c = new Car(4481, 1746, 1526);
6          System.out.println(a);
7          System.out.println(s);
8          System.out.println(c);
9      }
10 }
11
12 class Car {
13     private int length;
14     private int width;
15     private int height;
16
17     Car(int length, int width, int height) {
18         this.length = length;
19         this.width = width;
20         this.height = height;
21     }
22
23     String getOverallSize() {
24         return length + "×" + width + "×" + height;
25     }
26
27     void honk() {
28         System.out.println("嘀嘀");
29     }
30
31 }
```

源文件3-6　ObjectByString.java

```
1  public class ObjectByString {
2      public static void main(String[] args) {
3          int a = 45;
4          String s = "ABC";
```

```
 5       Car c = new Car(4481, 1746, 1526);
 6       System.out.println(a);
 7       System.out.println(s);
 8       System.out.println(c);
 9   }
10 }
11
12 class Car {
13     private int length;
14     private int width;
15     private int height;
16
17     Car(int length, int width, int height) {
18         this.length = length;
19         this.width = width;
20         this.height = height;
21     }
22
23     String getOverallSize() {
24         return length + "×" + width + "×" + height;
25     }
26
27     void honk() {
28         System.out.println("嘀嘀");
29     }
30
31     public String toString() {
32         return "Car[ length: " + length + ", width: " + width
             + ", height: " + width + "]";
33     }
34
35 }
```

由于几乎所有对象都需要输出到标准输出设备，所以通常会在自定义的类中提供 toString()方法来返回对象的字符串表示。而 JDK 中的类，如 String 类，则预定义了 toString 方法。

汽车类 Car 是对现实世界实体的抽象，称为实体类。实体类一般声明为 public,并单独保存在一个.java 源文件中。实体类的所有成员变量声明为 private,针对每个私有成员变量都声明一对 public 的 set()和 get()方法用来访问私有成员变量,这些方法称为 getter 和 setter。例如在源文件 3-7 中第 24 行至第 29 行声明了用于访问私有成员变量 length 的 getter 和 setter。第 5 行使用 setter 设置了成员变量的值为 4567,第 6 行通过 getter 读取变量的值。程序的输出如下：

```
Car[length: 4481, width: 1746, height: 1746]
4567
```

getter 和 setter 方法的名字约定为 getXxx 和 setXxx,其中 Xxx 是成员变量的名字,首字母大写。当成员变量是 boolean 类型时,getter 方法的名字约定为 isXxx。比如假设有 boolean 类型的成员变量 married,那么其 getter 方法的头部就是 boolean isMarriaged()。

实体类的构造方法一般有两个：无参数的构造方法(第 20 行)和全参数的构造方法(第 15 行)。"全参数"意思是为每个成员变量指定实际参数。无参数的构造方法通过 this 保留字调用了全参数的构造方法。

源文件 3-7　GetterSetter.java

```java
1  public class GetterSetter {
2      public static void main(String[] args) {
3          Car c = new Car(4481, 1746, 1526);
4          System.out.println(c);
5          c.setLength(4567);
6          System.out.println(c.getLength());
7      }
8  }
9
10 class Car {
11     private int length;
12     private int width;
13     private int height;
14
15     Car(int length, int width, int height) {
16         this.length = length;
17         this.width = width;
18         this.height = height;
19     }
20     Car() {
21         this(0, 0, 0);
22     }
23
24     public int getLength(){
25         return length;
26     }
27     public void setLength(int length){
28         this.length = length;
29     }
30
31     String getOverallSize() {
32         return length + "×" + width + "×" + height;
33     }
34
35     void honk() {
36         System.out.println("嘀嘀");
37     }
38
39     public String toString() {
40         return "Car[ length: " + length + ", width: " + width
               + ", height: " + width + "]";
41     }
42
43 }
```

3.5　方法的调用和返回

调用方法会把控制流转移到被调用的方法体中执行，执行完毕则返回到调用处继续执行后续语句。在进行方法调用时，需要把实际参数传递给形式参数。Java 语言仅有"传值"

一种方式：把实际参数的值赋值给形式参数。例如源文件3-8中声明了汽车类Car和司机类Driver,在第3行创建一个司机对象;在第4行使用司机对象创建了一个汽车对象并调用汽车对象的setSpeed(60)方法把车速设置为60。当程序执行第5行时,控制转移到第29行,同时把实际参数60赋值给形式参数speed,然后再去执行方法体,把形式参数的值赋值给成员变量。调用setSpeed(60)方法前和该方法执行完后汽车对象的状态如图3-5和图3-6所示。

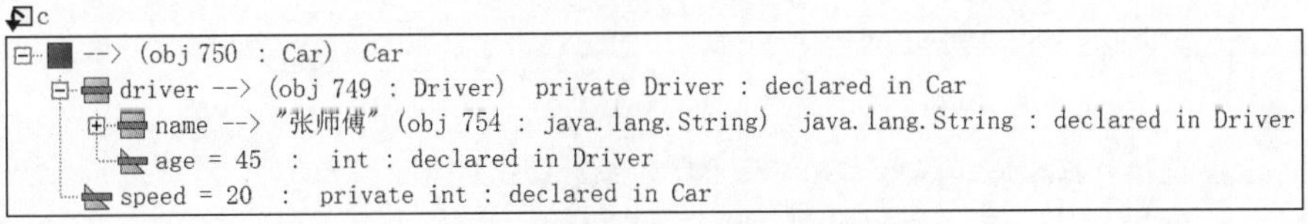

```
c
  --> (obj 750 : Car)    Car
     driver --> (obj 749 : Driver)    private Driver : declared in Car
        name --> "张师傅" (obj 754 : java.lang.String)   java.lang.String : declared in Driver
        age = 45      :  int : declared in Driver
     speed = 20    :  private int : declared in Car
```

图3-5 调用setSpeed(60)前的对象状态

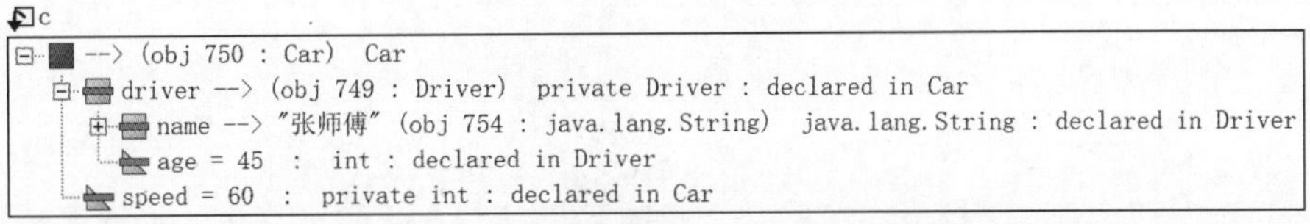

```
c
  --> (obj 750 : Car)    Car
     driver --> (obj 749 : Driver)    private Driver : declared in Car
        name --> "张师傅" (obj 754 : java.lang.String)   java.lang.String : declared in Driver
        age = 45      :  int : declared in Driver
     speed = 60    :  private int : declared in Car
```

图3-6 setSpeed(60)执行后的对象状态

在调用方法时给出的参数称为实际参数(arguments);在声明方法时列出的参数称为形式参数(parameters)。实际参数赋值给形式参数的过程称为形实结合。传值(passed by value)的形实结合方式就是把实际参数的值复制一份存放到形式参数中。形式参数与局部变量一样,当方法执行完毕,返回到调用点时就消失了。当形式参数有多个时,实际参数的类型和顺序必须与之匹配。

方法调用的一般过程如下。

(1) 按照从左到右的次序对实际参数计值。允许实际参数是个表达式,所以在进行形实结合之前,首先计算出表达式的值。

(2) 在栈区分配一块临时的存储区域,用以存储形式参数、局部变量、调用语句下一条语句的地址、当前寄存器信息,以及其他执行方法所需的工作单元等。

(3) 参数传递。把实际参数的值赋值给相应的形式参数。

(4) 执行方法体。

(5) 从方法中返回。当执行了方法体中的return语句,或者执行到了方法体的末尾,释放调用时分配的栈中存储局部变量的空间,控制返回到调用语句的下一条语句处执行。如果方法声明了返回值,则把返回值放在调用者与被调用者约定的存储单元中。

源文件3-8 **PassingByValue.java**

```
1 public class PassingByValue {
2     public static void main(String[] args) {
3         Driver zhang = new Driver("张师傅", 45);
4         Car c = new Car(zhang, 20);
5         c.setSpeed(60);
6         System.out.println(c.getSpeed());
7         zhang.setAge(46);
8         System.out.println((c.getDriver()).getAge());
```

```java
9      }
10 }
11
12 class Car {
13     private Driver driver;
14     private int speed;
15
16     Car(Driver driver, int speed) {
17         this.driver = driver;
18         this.speed = speed;
19     }
20
21     public void setDriver(Driver driver) {
22         this.driver = driver;
23     }
24
25     public Driver getDriver() {
26         return driver;
27     }
28
29     public void setSpeed(int speed) {
30         this.speed = speed;
31     }
32
33     public int getSpeed() {
34         return speed;
35     }
36
37     public void honk() {
38         System.out.println("嘀嘀");
39     }
40 }
41
42 class Driver {
43     String name;
44     int age;
45
46     Driver(String name, int age) {
47         this.name = name;
48         this.age = age;
49     }
50
51     void setName(String name) {
52         this.name = name;
53     }
54
55     String getName() {
56         return name;
57     }
58
59     void setAge(int age) {
60         this.age = age;
61     }
62
```

```
63    int getAge() {
64        return age;
65    }
66 }
```

形式参数的数据类型可以是基本数据类型,也可以是引用类型。当形式参数是引用类型时,传递的是对象引用而不是对象本身。例如源文件 3-8 第 4 行把引用变量 zhang 作实际参数调用了构造方法。在执行第 7 行前的司机对象和汽车对象状态如图 3-7 所示。注意,变量 zhang 和汽车对象中的成员变量 driver 引用的是同一个对象。

第 7 行把司机对象的年龄改为 46,该行执行后对象的状态如图 3-8 所示。当通过引用变量 zhang 把司机对象的年龄改为 46 后,汽车对象中的成员变量 driver 也知道了:因为二者引用的是同一个对象。

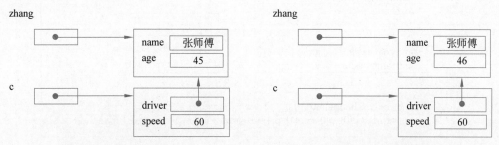

图 3-7 司机的年龄 45   图 3-8 司机的年龄 46

由于形式参数与局部变量一样,所以对形式参数的任何改变仅仅在方法体执行期间有效。一旦方法执行完毕,形式参数和局部变量一样,随着临时存储空间的释放而消失了。所以,对应基本数据类型的形式参数,方法体中对其做的任何改变,都将随着方法的返回而丢失。而对于引用类型的形式参数,由于形式参数仅仅从实际参数得到一份对象的引用,使得在方法执行期间,实际参数和形式参数同时引用了同一个对象。即便形式参数随着方法的返回而消失,调用者仍然可以通过实际参数访问对象。

如果在方法执行期间修改了形式参数所引用的对象,那么当方法执行结束后,就可以通过实际参数访问所做的修改。

源文件 3-9 声明了两个方法交换数组中两个元素的值:swap()方法和 exchange()方法。前者直接传递两个元素的值;后者则传递数组对象。当程序执行完毕会发现 swap()不能实现数组对象 a 中两个元素值的交换;而 exchange 则成功实现了交换。

源文件 3-9 ArrayAsParameter.java

```
1 public class ArrayAsParameter {
2     public static void main(String[] args) {
3         int[] a = {2, 3};
4         swap(a[0], a[1]);
5         exchange(a);
6     }
7
8     static void swap(int x, int y) {
9         int temp = x; x = y; y = temp;
10    }
11
```

```
12    static void exchange(int[] p) {
13        int temp = p[0]; p[0] = p[1]; p[1] = temp;
14    }
15 }
```

这是因为 swap 方法交换的是形式参数 x 和 y，而形式参数 x 和 y 随着 swap 的执行结束而消失了，数组对象仍然是原来那个数组对象，如图 3-9 和图 3-10 所示。而 exchange 则是通过形式参数 p 引用修改了数组对象，从而即使引用变量 p 消失了，通过引用变量 a 仍然能够访问被修改的数组对象，如图 3-11 和图 3-12 所示。

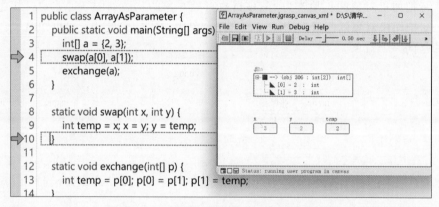

图 3-9　swap 方法返回之前的数组对象状态

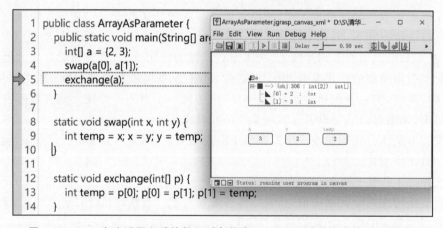

图 3-10　swap 方法返回之后的数组对象状态（x,y,temp 成为灰色表示不可用）

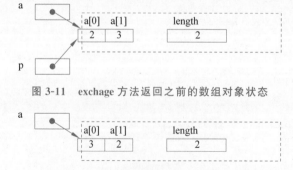

图 3-11　exchage 方法返回之前的数组对象状态

图 3-12　exchage 方法返回之后的数组对象状态

当执行了方法体中的 return 语句,或者方法体中的所有语句全部执行完毕,即执行到了方法体的末尾时,释放栈中的临时存储区域,控制返回到调用语句的下一条语句处执行。

用于从一个无返回值的方法中返回调用者的 return 语句形式如下:

```
return;
```

如果方法的返回值类型声明为 void,那么执行完所有语句后返回,也可以在方法体任意位置显式地通过 return 语句返回。如果方法声明需要返回值,那么就把返回值放在 return 保留字的后面:

```
return <返回值>;
```

<返回值>必须和方法所声明的返回值类型兼容。方法不仅能够返回一个基本类型的值,也能返回一个对象引用。在源文件 3-8 的第 33 行的 getSpeed()方法则返回一个整数;而第 25 行 getDriver()就是返回了司机对象的引用。

3.6 重载

Java 语言允许在同一个类中声明具有相同名字的方法,但这些方法的签名必须不同。这种现象称为方法重载(overloading)。重载现象在 JDK 的 API 中大量出现。例如 PrintWriter 类就有很多重载的 println()方法:

```
java.io.PrintWriter.println()
java.io.PrintWriter.println(boolean)
java.io.PrintWriter.println(char)
java.io.PrintWriter.println(char[])
java.io.PrintWriter.println(double)
java.io.PrintWriter.println(float)
java.io.PrintWriter.println(int)
java.io.PrintWriter.println(java.lang.Object)
java.io.PrintWriter.println(java.lang.String)
java.io.PrintWriter.println(long)
```

System.out 的数据类型就是 java.io.PrintWriter,即 System.out 引用了一个 java.io.PrintWriter 对象。之所以能够使用 System.out.println()方法输出任意数据类型的数据,就是得益于方法重载。方法重载使得语言的表达力更强,程序更简洁。

重载的方法是在一个类里面名字相同但参数类型序列不同的方法。方法返回类型可以相同也可以不同。每个重载的成员方法或者构造方法都必须有一个类内唯一的方法签名。

在源文件 3-10 的第 3 行和第 5 行声明了方法签名为 f(int)和 f(int, int)的方法,二者方法名都是 f,但一个参数类型序列是 int;另一个参数类型序列是 int, int,两个序列不同,所以两个方法签名不同,因此第 5 行的方法重载了第 3 行的方法。

与第 5 行声明的方法相比,第 8 行声明的方法虽然与其名字相同,返回值类型不同,但是方法签名是 f(int, int),方法签名相同,此处会出现编译错误消息:

```
OverloadDemo.java:8: error: method f(int,int) is already defined in
class OverloadDemo
    int f(int x, int y) {
```

第11行声明的方法尽管形式参数名与第5行的方法不同,但方法签名仍然相同,所以仍然造成编译错误:

```
OverloadDemo.java:11: error: method f(int,int) is already defined in
class OverloadDemo
    float f(int m, int n) {
```

第14行声明的方法的方法签名为f(int, float),与前面方法的方法签名不同,所以是一个重载方法。

第20行的方法g与第17行的方法g虽然访问修饰符不同,但方法签名相同,所以编译时会出现编译错误信息:

```
OverloadDemo.java:20: error: method g() is already defined in class OverloadDemo
    private void g() {
```

源文件 3-10 OverloadDemo.java

```
1  public class OverloadDemo {
2      void f(int a) {
3      }
4
5      void f(int a, int b) {
6      }
7
8      int f(int a, int b) {
9      }
10
11     float f(int m, int n) {
12     }
13
14     void f(int m, float n) {
15     }
16
17     public void g() {
18     }
19
20     private void g() {
21     }
22  }
```

3.7 类变量和实例变量

Java通过对象协作完成工作,在开始协作之前,首先根据类创建对象。那么,创建对象之前,就得把类装载到内存。当然,类只需装载一次。即下次使用该类时,如果已经装载,则不必重新装载。

如果类的成员变量特别声明在装载类时为其分配存储而不是创建对象时为其分配存储，则这些成员变量就称为静态变量（static variable）或者类变量（class variable）。类变量随着类的加载而产生，当程序运行结束才消亡。而那些创建对象时才分配存储的变量称为实例变量（instance variable）。实例变量随着某个对象创建而产生，随着该对象的消亡而消亡。在类的成员变量前如果有 static 保留字则声明该变量是静态变量，无 static 保留字就是实例变量。

当需要在整个应用程序范围共享某个变量时，可以声明为静态成员变量。例如：

源文件 3-11　Global.java

```
public class Global {
    public static int MAX = 1024;
    public static double PI = 3.141592653589793;
}
```

从静态成员变量所属类的成员方法中访问这些静态变量，直接使用变量名字。而从静态成员变量所属类的外面访问这些静态变量，则需要在变量前面增加类名和成员访问运算符"."。用法如下：

<类名>.<静态变量>

例如在源文件 3-12 中访问了类 Global 中的静态变量 PI，类 Global 定义在源文件 3-11 中。

源文件 3-12　StaticVarDemo.java

```
1 public class StaticVarDemo {
2     public static void main(String[] args)  {
3         System.out.println(Global.PI);
4     }
5 }
```

有两种方式初始化静态变量：声明时初始化和静态初始化块。例如，源文件 3-11 中，Global 类在声明静态变量 MAX 时就可以将其初值设置为 1024。静态块（static block）是类体中使用 static 保留字修饰的代码块。静态块仅在类装入时执行一次，通常对静态变量进行初始化。

下面的类 Global 在静态块中把静态变量 MAX 初值设置为 1024：

```
public class Global{
    public static int MAX;
    static {
        MAX = 1024;
    }
}
```

没有使用 static 关键字修饰的成员变量称为实例变量，通过对象引用访问实例变量。实例变量初始化有三种途径：声明时初始化、对象创建时初始化和通过方法调用初始化。

下面的例子声明了实例变量 aMember，同时将其初值设置为 1。这种初始化方式称为"声明时初始化"。

```
public class A {
    private int aMember = 1;
    public void aMethod() {
    }
}
```

下面的类 A 就使用构造方法把实例变量初始化为指定的值：

```
class A {
    private int aMember;
    A(int i) {
        aMember = i;
    }
}
```

使用成员方法也能进行实例变量的初始化：

```
class A {
    int aMember;
    void aMethod(int i) {
        aMember = i;
    }
}
```

总而言之，静态变量在类的存储空间中，不属于任何对象；而实例变量在对象的存储空间中，每个对象都有自己的实例变量。实例变量属于某个对象，必须创建了实例对象，其中的实例变量才会被分配空间，才能访问这个实例变量；静态变量不属于任何对象，而仅属于类，所以也称为类变量，只要程序加载了类的字节码，不用创建任何实例对象，静态变量就会被分配空间，静态变量就可以被访问了。实例变量必须创建对象后才可以通过这个对象的引用变量来访问，静态变量则可以直接使用类名来访问。

静态变量、实例变量和局部变量是 Java 程序中常见的三种变量。这些变量的生命周期不同：静态变量随着类的加载而产生，随着类的回收而消失；实例变量随着对象的创建而产生，随着对象的回收而消失。在方法体中声明的局部变量则随着方法的调用而出现在栈中，随着方法的返回而消失。方法的形式参数属于局部变量。

静态变量和实例变量都是成员变量，具有默认初值。局部变量没有默认初值，必须显式初始化。

在块中可以声明变量，这些变量也称为局部变量，仅能被该块中的语句访问。块(block)是花括号{}括起来的语句序列。在源文件 3-13 中第 4 行到第 6 行是一个块，其中声明了变量 b，该变量局部于其所在的块。第 10 行试图访问变量 b，则会出现编译错误：

```
BlockDemo.java:10: error: cannot find symbol
        System.out.println(b);
                           ^
    symbol:   variable b
    location: class BlockDemo
1 error
```

源文件 3-13　BlockDemo.java

```
1 class BlockDemo {
2     static int a;
3
4     {
5         int b = 0;
6     }
7
```

```
8    public static void main(String[] args) {
9        System.out.println(a);
10       System.out.println(b);
11   }
12 }
```

除了第 4 行到第 6 行的块,从第 8 行到第 11 行也是一个块,从第 1 行到第 12 行也是一个块。所以源文件 3-13 中共有 3 个块。块可以嵌套,但不可以交错。可以访问某个变量的块形成了该变量的作用域。一个变量在其声明的块 B 中总是可以被访问,在块 B 中也能访问外层块中的变量。在源文件 3-14 中,由于第 4 行声明的变量 i 局部于 main 块(即 main 的方法体),所以在同一块中的第 10 行访问 i 没有问题。第 11 行访问的 x 是在第 2 行声明的 x,而不是在第 6 行 for 循环体中声明的 x,因为后者仅作用于 for 循环体;而前者的作用域是类体,含类体中的所有块。第 17 行访问的 x 是在第 16 行声明的 x,而不是在第 2 行声明的 x,这种现象称为"就近原则"。程序的输出如下:

```
x in the for loop: 5
x in the for loop: 5
x in the for loop: 5
i is now: 3
x in the class: 2
x in aMethod: 1
```

源文件 3-14   ScopeOfVariable.java

```
1 public class ScopeOfVariable {
2     static int x = 2;
3     public static void main(String[] args) {
4         int i;
5         for (i = 0; i < 3; i++) {
6             int x = 0;
7             x = x + 5;
8             System.out.println("x in the for loop: " + x);
9         }
10        System.out.println("i is now: " + i);
11        System.out.println("x in the class: " + x);
12        aMethod();
13    }
14
15    static void aMethod() {
16        int x = 1;
17        System.out.println("x in aMethod: " + x);
18    }
19 }
```

第 6 行在 for 循环语句的循环体中声明了局部变量 x,那么这个 x 就与第 2 行声明成员变量 x 是两个不同的变量。按照"就近"原则,该循环体访问 x 的语句访问的是局部变量 x,而不是成员变量 x。由于当进入块的时候才创建局部变量,所以在 3 次循环中的每次循环都创建局部变量 x。当块执行完毕,该块中的局部变量消失,所以 3 次循环中的每次循环结束,x 都消失。

3.8 静态方法和实例方法

如果在成员变量的声明中增加了 static 修饰符,那么该方法就被声明为静态方法,否则称为实例方法。与通过对象引用调用实例方法不同,在静态方法所属类之外调用该静态方法必须使用类名和成员访问运算符"."。用法如下:

<类名>.<静态方法名>(<实际参数序列>)

如果一个方法所完成的功能与对象状态无关,则适合设计成为静态方法。例如下面的类 Thermograph 声明了从摄氏度(Centigrade)转换为华氏度(Fahrenheit)的方法,如果把摄氏度设计为方法的参数,则该方法所完成的转换与具体对象无关。因此设计为在 Thermograph 类的静态方法 centigradeToFahrenheit,如源文件 3-15 所示。

源文件 3-15　Thermograph.java

```
1 public class Thermograph {
2     public static int centigradeToFahrenheit(int cent) {
3         return cent * 9 / 5 + 32;
4     }
5 }
```

源文件 3-16 中的第 3 行调用该静态方法把 36℃转化为华氏度数:

源文件 3-16　StaticMethodDemo.java

```
1 public class StaticMethodDemo {
2     public static void main(String[] args){
3         System.out.println(Thermograph.centigradeToFahrenheit(36));
4     }
5 }
```

如果客户代码在静态方法所属的类中,则可以省略类名:

```
public class Thermograph{
    public static int centigradeToFahrenheit(int cent){
        return cent * 9 / 5 + 32;
    }
    public static int aMethod(){
        centigradeToFahrenheit(36);
    }
}
```

在同一个类中,静态方法只能访问静态变量;实例方法既能访问实例变量,也能访问静态变量。同样,在同一个类中,静态方法只能调用静态方法;实例方法既能调用实例方法,也能调用静态方法。于是,源文件 3-17 中存在两个编译错误:

i 是实例变量,而 main()方法是静态方法,由于静态方法只能访问同一个类中的静态变量,所以语句"int j=i"存在编译错误。

anInstanceMethod()声明为实例方法,不能被同一个类中的静态方法 main()调用。所以 main()方法体中的第二条语句也存在编译错误。

源文件 3-17　StaticMember.java

```java
public class StaticMember {
    int i = 8;
    static int k = 5;

    public static void main(String[] args) {
        int j = i;                    //由于 i 是实例变量,在静态方法 main 中不能直接访问
        anInstanceMethod();           //在静态方法中不能直接调用实例方法
        StaticMember sm = new StaticMember();
        sm.anInstanceMethod();
    }

    public void anInstanceMethod() {
        i = aStaticMethod();
    }

    public static int aStaticMethod() {
        return k;
    }
}
```

总之,静态方法从属于类,并不属于某个对象。静态方法不能访问其所在类的非静态成员。从静态方法所在类之外访问静态方法,需通过<类名>.<方法名>()的格式访问。

main 方法只能声明为静态的方法,因为它是整个 Java 应用的入口,发生在任何实例化之前。

3.9　命令行参数

在命令行提示符窗口运行 Java 程序时也可以给出参数,这些参数称为命令行参数。假设有 Arithmetic 类用来完成加法运算和减法运算。其 main 方法从命令行接收参数,然后输出计算结果。如图 3-13 所示,在命令行输入:

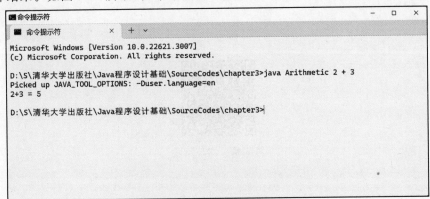

图 3-13　命令行参数

```
java Arithmetic 2 + 3
```

那么程序输出：

```
2 + 3 = 5
```

命令行参数通过 main 方法的形式参数 String[] args 传递。在这个例子中，命令行使用空白符作为分隔符给出了 3 个参数：2，+ 和 3。这三个实际参数分别传递给了 args[0]、args[1]和 args[2]。在源文件 3-18 第 13 行和第 14 行把两个操作数 args[0]和 args[2]转换为整数，第 17 行根据 args[1]执行加法或减法运算。

注意，当处理命令行参数时，一般需要检查实际参数是否与程序的期望一致。例如第 7 行检查实际参数的个数是否为 3 个。

源文件 3-18 Arithmetic.java

```
1  public class Arithmetic {
2      public static void main (String[] args) {
3          int operand1, operand2;
4          int result = 0;
5  
6          //检查命令行是否有 3 个参数
7          if (args.length != 3) {
8              System.err.println("Usage: java Arithmetic int  operator int");
9              return;
10         }
11  
12         //把字符串类型的参数转换为 int 类型
13         operand1 = Integer.parseInt(args[0]);
14         operand2 = Integer.parseInt(args[2]);
15  
16         //把第 2 个参数的首字符作为运算符
17         switch(args[1]) {
18             case "+" -> result = operand1 + operand2;
19             case "-" -> result = operand1 - operand2;
20             default ->{System.err.println("Error: invalid operator!");
21                        return;}
22         }
23         System.out.println(args[0]  + args[1] + args[2] + " = " + result);
24     }
25 }
```

第 3 章　章节测验

第 4 章 继 承

继承描述了类与类之间的"Is-a"关系。比如,轿车是机动车(A car is a vehicle)。假设类 B 继承了类 A,那么类 B 是类 A 的子类(child class),也称扩展类(extended class)或导出类(derived class)。反过来,类 A 是类 B 的父类(parent class),也称超类(super class)或基类(base class)。Java 语言中一个类只能有一个父类。

图 4-1 描述了类 A、B、C 等 8 个类之间的继承关系。其中,矩形框表示类;末端为空心三角形的箭头表示继承关系。如果从 B 到 A 有箭头,则表示 B 是 A 的子类。

这些具有继承关系的类就形成了一棵树,称为继承树。每个类就是树中的一个节点。

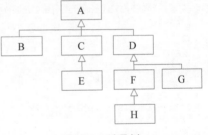

图 4-1 继承树

在这棵继承树中,具有相同父类的类称为兄弟类,如 B、C、D 互为兄弟。从树的根到某节点父类的路径上的所有类称为该节点的祖先类,即父类、父类的父类,直到根类。如类 H 的祖先类有 A、D、F。从树的根到某节点的路径称为继承链。以某节点为根的子树上的节点称为该节点的后代类。例如 D 的后代类有 F、G、H。

4.1 通过继承共享祖先的特征

根据中国公共安全行业标准 GA802—2014《机动车类型 术语和定义》,机动车(motor vehicle)是以动力装置驱动或者牵引,上道路行驶的供人员乘用或者用于运送物品以及进行工程专项作业的轮式车辆,包括汽车、摩托车、轮式专用机械车、挂车、有轨电车、特型机动车和上道路行驶的拖拉机。汽车是由动力驱动,具有 4 个或 4 个以上车轮的非轨道承载的车辆。假设有汽车类 Car 和机动车类 Vehicle 的声明如源文件 4-1 所示。

源文件 4-1　Car.java

```java
1 public class Car {
2     private String licenceNumber;           //车牌号码
3     private int length, width, height;      //外廓尺寸：长、宽、高
4 
5     public Car(String licenceNumber, int length, int width, int height) {
6         this.licenceNumber = licenceNumber;
7         this.length = length;
8         this.width = width;
9         this.height = height;
10    }
11
12    public String getLicenceNumber() {
13        return licenceNumber;
14    }
15
16    public void setLicenceNumber(String licenceNumber) {
17        this.licenceNumber = licenceNumber;
18    }
19
20    public String getOverallSize() {
21        return length + "×" + width + "×" + height;
22    }
23
24    public void honk() {
25        System.out.println("嘀嘀");
26    }
27
28 }
```

源文件 4-2　Vehicle.java

```java
1 public class Vehicle {
2     int length;
3     int width;
4     int height;
5
6     public Vehicle(int length, int width, int height) {
7         this.length = length;
8         this.width = width;
9         this.height = height;
10    }
11
12    public Vehicle() {
13        this(0, 0, 0);
14    }
15
16    public String getOverallSize() {
17        return length + "×" + width + "×" + height;
18    }
19
20 }
```

任何一辆机动车都有外廓尺寸属性，但不一定有车牌号码；任何一辆机动车都能被查看

外廓尺寸(getOverallSize),但不一定能够鸣笛(honk)。汽车是一种机动车,自然具有机动车的属性和行为。但汽车还有自己特有的属性和行为,比如车牌号码和设置车牌号码(setLicenceNumber)、查看车牌号码(getLicenceNumber)和鸣笛(honk)等。Java 语言中继承让子类自动具有父类的属性和行为。关键字 extends 用于声明类之间的继承关系,如源文件 4-3 所示。

源文件 4-3　ExtendsDemo.java

```
1 public class ExtendsDemo {
2     public static void main(String[] args) {
3         Car c = new Car("1234", 4500, 3400, 1600);
4         System.out.println(c.getOverallSize());
5     }
6 }
7 class Car extends Vehicle {
8     private String licenceNumber; //车牌号码
9
10     public Car(String licenceNumber, int length, int width, int height) {
11         super(length, width, height);
12         this.licenceNumber = licenceNumber;
13     }
14
15     public String getLicenceNumber() {
16         return licenceNumber;
17     }
18
19     public void setLicenceNumber(String licenceNumber) {
20         this.licenceNumber = licenceNumber;
21     }
22
23     public void honk() {
24         System.out.println("嘀嘀");
25     }
26
27 }
```

关键字 extends 声明类 Car 继承了类 Vehicle。这就使得类 Car 自动具有了 Vehicle 的属性外廓尺寸,以及行为查看外廓尺寸 getOverallSize()。所以在第 4 行就能够通过汽车对象的引用 c 来调用方法 getOverallSize(),尽管该方法并没有在 Car 类中声明。所以源文件 4-3 能够输出:

4500×3400×1600

子类不能继承父类的私有成员。源文件 4-2 中的机动车类 Vehicle 中的成员都是包私有的,可以被子类继承。

从子类实例化的对象中既有自身声明的成员变量,也有从父类继承的成员变量,如图 4-2 所示。

java.lang.Object 是所有类的父类,Object 类是唯一没有父类的类,是根类。Object 类定义了所有对象都应具备的行为,例如相等比较 equals(),

| 从父类继承的成员变量 |
| 从父类继承的成员方法 |
| 子类中声明的成员变量 |
| 子类中声明的成员方法 |

图 4-2　子类实例的成员

转换成字符串表示 toString()等。虽然类 Vehicle 没有显式地声明继承 Object 类,但 Java 语言默认已经继承了 Object 类。即下面显式继承 Object 类的声明:

```
public class Vehicle extends Object{
}
```

与 Vehicle 隐式继承 Object 类的声明等价:

```
public class Vehicle{
}
```

4.2 父类和子类的构造方法

子类不能继承父类的构造方法。但使用关键字 super 可以调用当前对象父类的构造方法。

在源文件 4-2 中,类 Vehicle 声明了构造方法 Vehicle(int length,int width,int height)。源文件 4-3 中的类 Car 在其构造方法 Car(String licenceNumber,int length,int width,int height)中的第一条语句(第11行)通过调用父类的构造方法来初始化继承来的成员变量。

如果在 Car 的构造方法中没有显式地调用有参数的父类的构造方法,那么编译器会在构造方法的第 1 行增加对父类无参构造方法的调用语句。在源文件 4-4 中第 10 行 Car 的构造方法中仅有一条访问子类中声明的成员变量的语句,没有对父类构造方法的调用语句,这种写法与

```
public Car(String licenceNumber, int length, int width, int height) {
    super();
    this.licenceNumber = licenceNumber;
}
```

等价。所以源文件 4-4 的输出是

```
0×0×0
```

而不是

```
4500×3400×1600
```

源文件 4-4 ConstructorDemo.java

```
1 public class ConstructorDemo {
2     public static void main(String[] args) {
3         Car c = new Car("1234", 4500, 3400, 1600);
4         System.out.println(c.getOverallSize());
5     }
6 }
7 class Car extends Vehicle {
8     private String licenceNumber;//车牌号码
9
10     public Car(String licenceNumber, int length, int width, int height) {
```

```
11            this.licenceNumber = licenceNumber;
12        }
13
14        public String getLicenceNumber() {
15            return licenceNumber;
16        }
17
18        public void setLicenceNumber(String licenceNumber) {
19            this.licenceNumber = licenceNumber;
20        }
21
22        public void honk() {
23            System.out.println("嘀嘀");
24        }
25
26 }
```

图 4-3 展示了默认的构造方法行为,即当在子类构造方法的第 1 行没有显式地调用父类构造方法时,默认调用父类的无参构造方法。

图 4-3 子类构造方法的默认行为

在子类的构造方法中调用父类的构造方法来实现对象初始化是很好的程序设计实践,因为这样做封闭了对父类的修改。例如,如果在父类中把成员变量的名字修改了,那么子类不用做任何修改。注意在子类构造方法中通过关键字 super 调用父类构造方法的语句必须是子类构造方法的首条语句。

如果在类的定义中没有定义任何构造方法,那么 Java 自动提供一个默认的无参构造方法。但是,一旦在类的定义中定义了构造方法,无论几个,那么 Java 就不再提供默认的无参构造方法。如果该类需要无参构造方法,那么必须显式地定义。所以在源文件 4-2 中 Vehicle 类声明了两个构造方法:一个是有参数的 Vehicle(int length, int width, int height),另一个是无参数的 Vehicle()。无参数的构造方法用于子类没有在构造方法中显式调用 Vehicle (int length, int width, int height)的情形。

如图 4-1 所示,A 是 C 的父类,C 是 E 的父类。假设声明如下:

```
class A {
}
class C extends A {
}
class E extends C {
}
```

该声明与下面的声明等价:

```
class A {
    A(){
        super();
```

```
        }
    }
    class C extends A {
        C(){
            super();
        }
    }
    class E extends C {
        E(){
            super();
        }
    }
```

那么,实例化 E 时会通过 super()调用 C 的无参构造方法 C();而在调用 C()时,就会通过 super()调用 A 的无参构造方法 A()。一般而言,实例化子类将会依次调用其继承链上的所有祖先类的构造方法。形成了构造方法链(constructor chaining)。源文件 4-5 展示了构造方法链。

源文件 4-5　ConstructorChain.java

```
1 public class ConstructorChain {
2     public static void main(String[] args) {
3         Child c = new Child();
4     }
5 }
6
7 class Ancestor {
8     Ancestor() {
9         System.out.println("Ancestor.");
10    }
11 }
12
13 class Parent extends Ancestor {
14     Parent() {
15         System.out.println("Parent.");
16    }
17 }
18
19 class Child extends Parent {
20     Child() {
21         System.out.println("Child.");
22    }
23 }
```

输出结果:

```
Ancestor.
Parent.
Child.
```

总而言之,从一个类创建对象会激发构造方法链上所有构造方法的执行。当使用子类的构造方法初始化对象时,良好的程序设计实践是在构造方法的第一条语句中使用含参数的 super 方法初始化父类的实例变量,然后再初始化子类的实例变量。

4.3 覆盖实例方法和隐藏静态方法

子类不仅能够继承和直接使用父类的实例方法,还能够使用原来的方法头部重新定义方法体。这种现象称为方法的覆盖(override)。toString 方法是类 java.lang.Object 中的实例方法,该方法返回对象的字符串表示。System.out.println(a) 实际上调用了 a 的 toString 方法,即 System.out.println(a.toString())。源文件 4-6 中的第 4 行试图输出变量 a 所引用对象的字符串表示,但是在类 A 中并没有定义约定的 toString 方法返回对象的字符串表示,那么就会执行类 A 的父类 Object 类中的 toString 方法,输出结果是:

```
A@28a418fc
```

其中,A 是对象的类名,符号@(读作 at)后面是该对象的内存地址。

源文件 4-6　OverrideDemo.java

```
1 public class OverrideDemo {
2     public static void main(String[] args) {
3         A a = new A();
4         System.out.println(a);
5     }
6 }
7
8 class A {
9     int aMember;
10 }
```

可以在类 A 中重新定义 toString 方法,如源文件 4-7 所示。输出结果就变成了:

```
A[aMember = 0]
```

在子类 A 覆盖了父类 Object 的实例方法 toString 的情况下,访问 A 的实例 a 上的 toString 就是调用类 A 中定义的 toString 方法。

源文件 4-7　OverrideToStringDemo.java

```
1 public class OverrideToStringDemo {
2     public static void main(String[] args) {
3         A a = new A();
4         System.out.println(a);
5     }
6 }
7
8 class A {
9     int aMember;
10     public String toString() {
11         return "A[aMember = " + aMember + "]";
12     }
13 }
```

方法的覆盖使得子类有机会修改父类的行为。需要注意以下 4 点。
(1) 子类的方法与父类中被覆盖方法的返回值类型、名字、参数类型序列一致。如果被覆盖方法的返回值类型为 void，覆盖方法的返回值类型也必须是 void；如果被覆盖方法的返回值类型为基本类型，覆盖方法的返回值类型也必须是该基本类型；如果被覆盖方法的返回值是引用类型，覆盖方法的返回值必须是与之兼容的引用类型。

（2）只有那些可继承的实例方法才可以被覆盖。也就是说，子类可以覆盖在同一包中父类的非私有实例方法，以及不同包中的父类的 public 或 protected 实例方法。

（3）子类中的方法可以保持或提高被覆盖方法的可访问性，但不能降低。可访问性从小到大依次是包私有（默认）、protected 和 public。

（4）对象的类型决定了所执行的实例方法的版本，而不是引用变量的类型。

Java 语言的注解 @Override 用来提示编译器这是一个覆盖方法，注意检查该方法是否满足方法覆盖的条件，如源文件 4-8 第 10 行所示。如果有 @Override 注解，而在其后写成了 public void toString()，编译器会指出错误而不认为这个 toString 是新定义的实例方法。

源文件 4-8　OverrideAnotationDemo.java

```
1  public class OverrideAnotationDemo {
2      public static void main(String[] args) {
3          A a = new A();
4          System.out.println(a);
5      }
6  }
7
8  class A {
9      int aMember;
10     @Override
11     public String toString() {
12         return "A [aMember = " + aMember + "]";
13     }
14 }
```

使用保留字 super 可以显式地访问被覆盖的实例方法。在源文件 4-9 中声明了 Vehicle 类，并在其中声明了 toString 方法。在第 17 行 Car 类覆盖该方法声明了自己的 toString 方法。第 18 行使用保留字 super 显式地调用了被覆盖的 Vehicle 类中声明的 toString 方法；否则，如果不使用 super，那么将调用 Car 类中声明的 toString 方法。程序的输出为：

```
licenceNumber: 1234, length: 4500, width: 3400, height: 1600
```

保留字 super 也可以用来区分子父类继承关系中同名的实例变量。

源文件 4-9　SuperDemo.java

```
1  public class SuperDemo {
2      public static void main(String[] args) {
3          Car c = new Car("1234", 4500, 3400, 1600);
4          System.out.println(c);
5      }
6  }
7
8  class Car extends Vehicle {
9      private String licenceNumber;        //车牌号码
10
11     Car(String licenceNumber, int length, int width, int height) {
```

```
12          super(length, width, height);
13          this.licenceNumber = licenceNumber;
14      }
15
16      @Override
17      public String toString() {
18          return "licenceNumber: " + licenceNumber + ", "+ super.toString();
19      }
20
21      public void honk() {
22          System.out.println("嘀嘀");
23      }
24
25  }
26
27  class Vehicle {
28      int length;
29      int width;
30      int height;
31
32      Vehicle(int length, int width, int height) {
33          this.length = length;
34          this.width = width;
35          this.height = height;
36      }
37
38      Vehicle() {
39          this(0, 0, 0);
40      }
41
42      public String toString() {
43          return "length: " + length + ", width: " + width + ", height: " + height;
44      }
45  }
```

如果在子类中改变了父类的静态方法的方法体,这种现象就称子类中的方法隐藏（hidden）了父类的方法。源文件 4-10 展示了方法的覆盖与隐藏特点。

源文件 4-10　HiddenDemo.java

```
1  public class HiddenDemo {
2      public static void main(String[] args) {
3          Child.staticMethodA();
4          Child.staticMethodB();
5      }
6  }
7
8  class Parent {
9      static void staticMethodA() {
10         System.out.println("The static method A in Parent.");
11     }
12
13     static void staticMethodB() {
14         staticMethodA();
```

```
15      }
16  }
17
18  class Child extends Parent {
19      static void staticMethodA() {
20          System.out.println("The static method A in Child.");
21      }
22  }
```

由于子类 Child 隐藏了父类的静态方法 staticMethodA()，所以 Child.staticMethodA() 调用的是子类中的方法。由于子类 Child 继承了父类的静态方法 staticMethodB()，所以 Child.staticMethodB() 调用的是父类中的方法 staticMethodB()，而此方法又调用了父类的静态方法 staticMethodA()。所以程序的输出结果是：

```
The static method A in Child.
The static method A in Parent.
```

4.4 上转型和下转型

在源文件 4-11 中声明了父类 Vehicle 和子类 Car。在第 4 行让变量 c 引用 Car 类型的对象，在第 7 行把变量 c 赋值给 Vehicle 类型的引用变量 v。v 用来引用 Vehicle 类型的对象；而 c 引用 Car 类型的对象。虽然变量 v 和 c 的类型不同，但是因为 Vehicle 是 Car 的父类，所以这样的赋值合法有效。按照某类的祖先类引用该类的对象称为上转型（upcasting）。为了上转型 Car 对象，只需要把该对象的引用赋值给 Vehicle 类型的变量即可。

当通过上转型引用子类的对象时，无法访问子类中定义的实例变量和非覆盖方法。例如在源文件 4-11 中的第 7 行把 Car 类的对象上转型为 Vehicle 类型，在第 8 行访问父类中未被覆盖的实例方法 getOverallSize 没有问题；在第 9 行访问父类中定义的实例变量 length 也没有问题。但是，第 10 行试图访问子类对象的实例变量 licenceNumber 却产生了编译错误：找不到符号 licenceNumber。类似地，在第 11 行试图调用子类对象的实例方法 honk，但产生编译错误：找不到符号 honk。

源文件 4-11　CastingDemo.java

```
1  public class CastingDemo {
2
3      public static void main(String[] args) {
4          Car c = new Car("1234", 4500, 3400, 1600);
5          System.out.println(c.getOverallSize());
6
7          Vehicle v = c;
8          System.out.println(v.getOverallSize());
9          System.out.println(v.length);
10         System.out.println(v.licenceNumber);    //错误：找不到符号
11         v.honk();                                //错误：找不到符号
```

```
12
13          Car a = (Car)v;
14          a.honk();
15
16          Car b;
17          if (v instanceof Car) {
18              b = (Car)v;
19              b.honk();
20          }
21      }
22 }
23
24 class Car extends Vehicle {
25      String licenceNumber;
26
27      Car(String licenceNumber, int length, int width, int height) {
28          super(length, width, height);
29          this.licenceNumber = licenceNumber;
30      }
31
32      void honk() {
33          System.out.println("嘀嘀");
34      }
35 }
36
37 class Vehicle {
38      int length;
39      int width;
40      int height;
41
42      Vehicle(int length, int width, int height) {
43          this.length = length;
44          this.width = width;
45          this.height = height;
46      }
47
48      Vehicle() {
49          this(0, 0, 0);
50      }
51
52      String getOverallSize() {
53          return length + "×" + width + "×" + height;
54      }
55 }
```

但如果在父类 Vehicle 中增加了实例方法 honk() 的定义(源文件 4-12 第 56 行到第 58 行)：

源文件 4-12　honk

```
42      Vehicle(int length, int width, int height) {
43          this.length = length;
44          this.width = width;
45          this.height = height;
46      }
```

```
47
48      Vehicle() {
49          this(0, 0, 0);
50      }
51
52      String getOverallSize() {
53          return length + "×" + width + "×" + height;
54      }
55
56      String honk() {
57          ;
58      }
59 }
```

这使得父类 Vehicle 中的 honk() 方法被覆盖,则不会在第 11 行产生编译错误。因为当前的对象是 Car 类型的对象,所以 v.honk() 调用的是子类 Car 中定义的实例方法 honk() 而不是父类中定义的实例方法 honk()。

如果沿着继承链从祖先类向子类方向转型,则称为下转型(downcasting)。上转型对象总是允许的,但下转型则需要显式地说明。通过在对象引用前使用转型运算符,(<类型>),来实现显式的转型。例如在源文件 4-11 中第 13 行:

```
Car a = (Car)v;
```

就是把按照父类 Vehicle 引用的对象转型为子类 Car 类型。下转型就意味着按照子类型访问父类型的对象,如果被转型的对象不是子类型的实例,显然存在风险。比如语句:

```
Vehicle v = new Vehicle();
Car c = (Car)v;
String x = c.licenceNumber;
```

而事实上 c 所引用的对象中根本就没有实例变量 licenceNumber。

所以为了保证转型成功,程序必须确保被下转型的对象是所转类型的实例。一般的做法是在转型之前对被转型的对象进行判定,看看是否是所转类型的实例。这就用到了运算符 instanceof。如果对象是某个类的实例,运算符 instanceof 则返回 true。源文件 4-11 中的第 17 行在下转型之前使用 instanceof 进行判定:

```
if (v instanceof Car){
    b = (Car)v;
    b.honk();
}
```

如果一个对象 o 是类型 C 的实例,或者 o 是 C 后代类的实例,那么 o instanceof C 返回 true;如果 o 是 null,instanceof 返回 false。

4.5 抽象类和抽象方法

如果想限制一个类不能实例化对象,那么就是要关键字 abstract 将其声明为抽象类(abstract class)。例如考虑机动车类 Vehicle。如果先实例化了一个轿车 Car 对象,然后将其上转型为机动车,因为轿车也是机动车,所以这种做法自然,也合理。但是,上转型后无法通过 Vehicle 类型的引用变量访问 Car 中声明的非覆盖实例方法。例如源文件 4-11 中的第 11 行所示。在 Vehicle 类中增加实例方法 honk() 的定义(源文件 4-12 第 56 行到第 58 行)能够解决第 11 行的错误,但是该方法体中只有一条空语句,因为"机动车"太抽象了,无法定义如何鸣笛。需要有一个实例方法,但不想或不能定义方法体,就可以把这种方法定义为抽象方法,含有抽象方法的类必须声明为抽象类。在声明 Vehicle 时,增加 abstract 关键字,如源文件 4-13 所示。

源文件 4-13　AbstractDemo.java

```
1 public class AbstractDemo {
2
3     public static void main(String[] args) {
4         Car c = new Car("1234", 4500, 3400, 1600);
5         c.honk();
6
7         Vehicle v = c;
8         v.honk();
9
10        v = new Vehicle();        //错误:Vehicle 是抽象类,不能被实例化
11    }
12 }
13
14 class Car extends Vehicle {
15     String licenceNumber;
16
17     Car(String licenceNumber, int length, int width, int height) {
18         super(length, width, height);
19         this.licenceNumber = licenceNumber;
20     }
21
22     String getOverallSize() {
23         return length + "×" + width + "×" + height;
24     }
25
26     void honk() {
27         System.out.println("嘀嘀");
28     }
29 }
30
31 abstract class Vehicle {
32     int length;
33     int width;
34     int height;
35
36     Vehicle(int length, int width, int height) {
37         this.length = length;
38         this.width = width;
39         this.height = height;
40     }
41
42     Vehicle() {
43         this(0, 0, 0);
```

```
44    }
45
46    abstract String getOverallSize();
47
48    abstract void honk();
49 }
```

如果一个方法只定义了方法的头部而没有定义方法体,这样的方法就称为抽象方法(abstract method)。源文件 4-13 的第 48 行就把实例方法 honk 声明为抽象方法:方法头部前面增加 abstract 关键字,方法头部使用分号(;)结尾。抽象方法中没有方法体。

含有抽象方法的类必须是抽象类,所以在第 31 行 Vehicle 类同时使用 abstract 关键字定义为抽象的。抽象类只能作为基类,不能被实例化。Vehicle 类的实例方法 getOverallSizeVehicle 也可以设计为抽象方法(第 46 行)。通过把 Vehicle 定义为抽象类禁止从该类实例化对象,从而避免调用抽象方法。子类应提供所有抽象方法的具体实现。如果子类没有全部实现父类中的抽象方法,则子类必须声明为抽象的。第 22 行和第 26 行的 Car 类分别提供了抽象方法 getOverallSizeVehicle 和 honk 的实现。抽象类中可以不含抽象方法。抽象类中可以含有构造方法用来供子类的构造方法调用以初始化父类实例变量。例如在第 18 行 Car 类的构造方法中就调用了在第 36 行声明的 Vehicle 类的构造方法。抽象类中的非抽象方法可被后代类共享。

4.6 保留字 final

保留字 final 的含义是"最终的,不可改变的"。用来修饰类、方法和变量。被 final 修饰的类不能被继承;被 final 修饰的方法不能被覆盖;被 final 修饰的变量不能被再次赋值,只能初始化时赋值一次。

前文介绍了类与类之间具有继承关系。如果想限制某个类不能被任何类所继承,那么就在定义类的时候使用 final 来修饰:

```
final class A {
    ……
}
```

假如类 B 试图继承类 A,那么就会出现编译错误:The type B cannot subclass the final class A(类 B 不能作为 final 类的子类)。一个类不能既是 final 类又是 abstract 类。

final 也可以修饰类的成员。被 final 修饰的类的成员方法不能被子类覆盖。被 final 修饰的类的成员变量一旦初始化不能再被修改。源文件 4-14 展示了 final 成员的特点。

源文件 4-14　FinalMemberDemo.java

```
1 public class FinalMemberDemo {
2
3     public static void main(String[] args) {
4         A a = new A();
5         a.f();
6     }
7 }
```

```
8
9 class A {
10     public static final double PI = 3.141592653589793;     //公共的静态常量
11     private static final double TAU = 2.0 * PI;            //私有的静态常量
12     private final int max = 2048;                          //私有的实例常量
13     final int[] a = {1, 2, 3};
14     static final int[] INDEX = {4, 5, 6};
15
16     final   void f() {
17         PI = 3.14;                                          //不能再次赋值
18         TAU = 6.28;                                         //不能再次赋值
19         max = 1024;                                         //不能再次赋值
20
21         a = new int[3];                                     //不能再次赋值
22         INDEX = a;                                          //不能再次赋值
23
24         a[0] = 10;
25         INDEX[0] = 40;
26     }
27 }
28
29 class B extends A {
30     public void f() {    //错误：B 中的 f()不能覆盖 A 中的 f()
31         a[1] = 20;
32     }
33 }
```

在源文件 4-14 中的类 A 中首先使用 final 声明了三个常量 PI、TAU 和 max，分别是公共静态、私有静态、私有实例常量，第 17、18、19 行的赋值产生编译错误。所以无论成员变量是否静态，只要被 final 修饰，就不能被再次赋值。

第 13 行声明并初始化了整型数组类型的引用变量 a，并引用数组{1,2,3}；第 14 行声明并初始化了整型数组类型的静态引用变量 INDEX，并引用数组{4,5,6}。引用类型的成员变量 a 和 INDEX 也被 final 修饰，方法 f()对其的修改企图也会产生编译错误。但注意，是变量 a 或 INDEX 的值不能被修改，而不是 a 或 INDEX 所引用的对象不能被修改。第 24 行和第 25 行修改了数组中某元素的值。数组对象的引用变量 a 被声明为 final，其含义是引用类型变量 a 的值不能被修改，但并不意味着 a 所引用的数组对象不能被修改。

类 B 继承了类 A，并试图覆盖方法 f()，则出现编译错误，因为类 A 中把方法 f 声明为 final，不允许被覆盖。

总之，如果一个类不允许被子类继承，就使用 final 修饰；如果一个方法不允许被子类覆盖，就用 final 修饰；如果一个成员变量不允许被再次赋值，就用 final 修饰。在 Java 应用中全局常量通常声明为 public static final XXX，其中 XXX 是常量名，大写。若由多个单词组成，则使用下画线连接，如 XXX_XXX。

4.7 接口

从形式上看，接口是一组抽象方法。从内容上看，接口是某功能的使用者与该功能实现者之间的一组约定。

假如一个应用程序想使用汽车对象,要求这个汽车对象提供鸣笛(honk)、查看外廓尺寸(getOverallSize)、查看车牌号码(getLicenceNumber)的功能,开发团队中指派小张负责设计实现小汽车对象,并告诉小张具体这些方法的头部:

```
String getLicenceNumber()
String getOverallSize()
void honk()
```

把这些方法头部声明在接口中,如源文件 4-15 所示。保留字 interface 声明接口,后面是接口的名字。在接口 ICar 中声明 3 个抽象方法。

源文件 4-15　ICar.java

```
1  public interface ICar {
2
3      /**
4       * 查看车牌号码
5       * @return 车牌号码,如果无车牌号码则返回 null
6       */
7      String getLicenceNumber();
8
9      /**
10      * 查看外廓尺寸
11      * @return 外廓尺寸,形如 M 长×宽×高,单位 mm. 例如 M4500×3400×1600
12      */
13     String getOverallSize();
14
15     /**
16      * 模拟鸣笛,在标准输出显示字符串 "嘀嘀"
17      * @return 无
18      */
19     void honk();
20 }
```

把接口 ICar 交给小张后,小张就应该做两件事情:设计接口的实现类和测试实现类。保留字 implements 用来建立接口与其实现类之间的实现关系。由于接口中的方法的可访问性默认是 public,所以其实现类中的相应方法也必须是 public 方法。假设小张设计的实现类如源文件 4-16 所示。

源文件 4-16　Car.java

```
1  public class Car implements ICar {
2      String licenceNumber;
3      int length;
4      int width;
5      int height;
6
7      Car(String licenceNumber, int length, int width, int height) {
8          this.licenceNumber = licenceNumber;
9          this.length = length;
10         this.width = width;
11         this.height = height;
12     }
13
```

```
14    public String getLicenceNumber() {
15        return licenceNumber;
16    }
17
18    public String getOverallSize() {
19        return length + "×" + width + "×" + height;
20    }
21
22    public void honk() {
23        System.out.println("嘀嘀");
24    }
25 }
```

设计好实现类后应该马上对实现类进行测试，以确认是否满足接口要求。源文件 4-17 在第 4 行声明了 ICar 类型的引用变量 c，创建了实现类 Car 类型的对象并将其引用赋值给 c。通常使用接口类型声明变量而不是使用实现类，因为一个接口可以有多个实现类。把变量声明为接口类型会使程序具有更好的可扩展性。第 5 行到第 7 行分别调用了接口中约定的方法，程序输出如下：

```
嘀嘀
1234
4500×3400×1600
```

可以看到，满足接口中的要求。

源文件 4-17   InterfaceDemo.java

```
1 public class InterfaceDemo {
2
3     public static void main(String[] args) {
4         ICar c = new Car("1234", 4500, 3400, 1600);
5         c.honk();
6         System.out.println(c.getLicenceNumber());
7         System.out.println(c.getOverallSize());
8     }
9 }
```

对类的测试称为单元测试。建议使用 jUnit 测试框架进行单元测试。源文件 4-18 是类 Car 的测试驱动程序，其中 @Before 注解 setUp 方法会在每个 @Test 注解的方法执行前被调用；@Test 注解的方法称为一个测试用例。测试用例由两部分组成：输入、预期的输出。输入以被测方法的实际参数提供。详细的基于 jUnit 的单元测试介绍见附录 C。

源文件 4-18   CarTest.java

```
1 import org.junit.Assert;
2 import org.junit.Before;
3 import org.junit.Test;
4
5 public class CarTest {
6     ICar c;
7     @Before public void setUp() {
8         c = new Car("1234", 4500, 3400, 1600);
9     }
10
```

```
11    @Test public void getLicenceNumberTest() {
12        Assert.assertEquals("1234", c.getLicenceNumber());
13    }
14
15    @Test public void getOverallSizeTest() {
16        Assert.assertEquals("4500×3400×1600", c.getOverallSize());
17    }
18 }
```

接口也可以继承接口,通过继承,子接口具有了父接口的成员。在 Java 中接口可以同时继承多个接口。源文件 4-19 到源文件 4-22 展示了接口的多重继承情况。

源文件 4-19   IA.java

```
1 public interface IA {
2     void f();
3 }
```

源文件 4-20   IB.java

```
1 public interface IB {
2     void g();
3 }
```

源文件 4-21   IC.java

```
1 public interface IC extends IA, IB{
2 }
```

源文件 4-22   T.java

```
1 public class T implements IC {
2     public void f() {
3         System.out.println("Do something of f");
4     }
5
6     public void g() {
7         System.out.println("Do something of g");
8     }
9 }
```

由于接口 C 既继承了接口 A,又继承了接口 B,所以接口 C 就具有了成员:抽象方法 f()和抽象方法 g()。因此类 T 必须提供这两个方法的实现。

一个类可实现多个接口。源文件 4-23 展示了类 S 同时实现了两个接口 IA 和 IB。

源文件 4-23   S.java

```
1 public class S implements IA, IB {
2     public void f() {
3         System.out.println("Do something of f");
4     }
5
6     public void g() {
7         System.out.println("Do something of g");
8     }
9 }
```

从以上实例可以总结出定义接口的语法：

```
interface <名字>{
    <接口成员>
}
```

除了抽象方法，接口的成员还可以是静态常量、默认方法(JDK 8)、静态方法(JDK 8)和私有方法(JDK 9)等。

定义接口的实现类的语法：

```
class <类名> implements <接口1>,…,<接口n> {
    //对接口中所有抽象方法的实现
    //其他类成员定义
}
```

实现某个接口意味着实现接口的所有抽象方法和继承接口中所有的静态常量。如果不想实现接口中的全部抽象方法，则必须把实现类声明为抽象类。

可访问性为 public 的接口与可访问性为 public 的类一样，其对应的源文件名必须与接口名严格一致。比如：

```
public interface IA {
    void f();
}
```

必须保存为 IA.java。

接口中的抽象方法默认使用 public 和 abstract 修饰。在接口中也可以定义成员变量，这些成员变量默认使用 public、static 和 final 修饰，且必须给出初值。

例如：

```
public interface ID {
    int CAPACITY = 128;
}
```

定义了静态常量 CAPACITY，初值为 128。等价于：

```
public interface ID {
    public final static int CAPACITY = 128;
}
```

访问接口中定义的常量与访问类的静态成员类似，通过"<接口名>.<常量名>"格式访问。例如：

```
public class IDDemo {
    public static void main(String[] args) {
        int[] a = new int[ID.CAPACITY];
    }
}
```

如果接口中某个方法在多个实现类中以相同方法实现，为了避免源代码重复，可以默认方法的形式声明在接口中。使用保留字 default 声明默认方法：

```
public default 返回类型 方法名 () {}
```

源文件 4-24 和源文件 4-25 中分别声明的接口 IE 及其实现类 E。在接口 IE 中声明了默认方法f。源文件4-26通过IE类型的变量引用了E类型的对象；并通过该引用调用了方法f。因为f是声明在接口IE中的默认方法，而E是接口IE的实现类，共享了该默认方法，相当于在E中声明了f，所以能够在第4行调用该方法。

源文件 4-24　IE.java

```
1 public interface IE {
2     public default void f () {
3         System.out.println("A default method.");
4     }
5 }
```

源文件 4-25　E.java

```
1 public class E implements IE {
2
3 }
```

源文件 4-26　DefaultDemo.java

```
1 public class DefaultDemo {
2     public static void main(String[] args) {
3         IE e = new E();
4         e.f();
5     }
6 }
```

如果父类中的实例方法和所实现的接口中的默认方法发生名字冲突，优先访问继承的实例方法。如果实现的多个接口中出现了同名的默认方法，那么实现类需要显式覆盖该默认方法。使用"<接口名>.super.<默认方法名>"调用被覆盖的默认方法。

在接口中可以使用 static 保留字声明静态方法：

```
public interface IF {
    public static void f () {
        System.out.println("接口中的静态方法");
    }
}
```

类似于访问类中声明的静态方法，使用<接口名>.<静态方法名>(<实参序列>)访问接口中的静态方法。

```
public class StaticMethodDemo {
    public static void main(String[] args) {
        IF.f();
    }
}
```

接口中声明的静态方法不能被子接口继承。

4.8　多态

多态(Polymorphism)是程序在运行时刻展现出的一种现象：相同的消息通过相同类型的引用发送给不同的对象却产生了不同的行为。比如，有 A、B、C 三辆机动车停在路边，其中 A 是轿车、B 是救护车、C 是警车。现在让每辆机动车都开始鸣笛。那么，这三辆车应当发出不同的鸣笛声，虽然都是机动车。这就是生活中的多态现象。

类似的现象，即通过相同类型的引用发送相同的消息，由于运行时接收并执行消息对象的类型不同，从而产生不同行为，在 Java 程序运行时刻也可以发生。

多态机制为 Java 程序设计带来了便利：允许把不同子类的对象统一看作其祖先类的对象进行组织，比如安排到线性表中，或者安排到树中，或者图中。当执行这些对象上的方法时，Java 运行时刻根据当时对象的类型选择该类型上定义的方法版本执行。

多态既可以由类继承实现，也可以通过接口及其实现类完成。源文件 4-27 展示了通过继承实现多态的程序设计。

源文件 4-27　PolyByInheritanceDemo.java

```
1  abstract class Vehicle {
2      /**
3       * 模拟鸣笛,在标准输出显示字符串"嘀嘀"
4       * @return 无
5       */
6      public abstract void honk();
7  }
8
9  class Car extends Vehicle {
10     public void honk() {
11         System.out.println("嘀嘀");
12     }
13 }
14
15 class FireTruck extends Vehicle {
16     public void honk() {
17         System.out.println("呜…");
18     }
19 }
20
21 class Ambulance extends Vehicle {
22     public void honk() {
23         System.out.println("嘀嘟嘀嘟");
24     }
25 }
26
27 public class PolyByInheritanceDemo {
28
29     public static void main(String[] args) {
30         Vehicle a, b, c;
31         a = new Car();
32         b = new FireTruck();
33         c = new Ambulance();
34         a.honk();
35         b.honk();
36         c.honk();
37     }
38 }
```

其输出结果是：

嘀嘀
呜…
嘀嘟嘀嘟

在源文件 4-27 中首先声明了父类 Vehicle，然后通过继承 Vehicle 声明了三个子类汽车 Car、救火车 FireTruck 和救护车 Ambulance。父类 Vehicle 中声明了一个抽象的鸣笛方法 honk，意思是机动车应该具有鸣笛行为，但如何鸣笛由具体的子类决定。三个子类分别提供了鸣笛 honk 的具体实现，即方法体。在第 30 行使用父类 Vehicle 类型的引用变量 a、b、c 分别引用了轿车、救火车和救护车三个对象，这三个对象的类型分别是 Car、FireTruck 和 Ambulance。

当分别调用 a、b、c 三个对象上的方法 honk() 时，由于此时 a、b、c 引用的三个对象分别是 Car、FireTruck 和 Ambulance 类型，Java 虚拟机就会把 honk 与此时的具体对象的实例方法 honk 绑定，所以分别调用了这些子类中定义的 honk() 方法。这样，虽然变量 a、b、c 的类型相同、接收的消息相同(a.honk()、b.honk()、c.honk())，但却产生不同的结果。这种现象就是多态现象。

图 4-4　不同机动车上的"鸣笛"行为

由于任意类的实例都可以上转型为其任意祖先类的类型，所以在类继承链上任意节点都可以设计多态程序。

源文件 4-28 展示了如何通过接口及其实现类设计多态程序。

源文件 4-28　PolybyInterfaceDemo.java

```
1  interface ICar {
2  
3      /**
4       * 模拟鸣笛,在标准输出显示字符串"嘀嘀"
5       * @return 无
6       */
7      void honk();
8  }
9  
10 class Car implements ICar {
11     public void honk() {
12         System.out.println("嘀嘀");
13     }
14 }
15 
16 
17 class FireTruck implements ICar {
```

```
18        public void honk() {
19            System.out.println("呜…");
20        }
21 }
22
23 class Ambulance implements ICar {
24        public void honk() {
25            System.out.println("嘀嘟嘀嘟");
26        }
27 }
28
29 public class PolybyInterfaceDemo {
30
31        public static void main(String[] args) {
32            ICar a, b, c;
33            a = new Car();
34            b = new FireTruck();
35            c = new Ambulance();
36            a.honk();
37            b.honk();
38            c.honk();
39        }
40 }
```

在源文件 4-28 中首先声明了接口 ICar，其中只有一个抽象方法 honk。然后声明了该接口的三个实现类：Car、FireTruck 和 Ambulance。在第 32 行使用接口声明了三个变量 a、b、c 分别用来引用 Car 对象、FireTruck 对象和 Ambulance 对象。最后向 a、b、c 发出相同的消息 honk，虽然这三个变量都是 ICar 类型，但在运行时刻引用了不同的对象，Java 虚拟机在运行时刻把消息 honk 与此刻对象的 honk 绑定，使得看起来像是向同类型的对象发送了相同的消息，但接收消息的对象却展现不同的行为。程序产生如下输出：

```
嘀嘀
呜…
嘀嘟嘀嘟
```

通过接口实现多态比通过继承实现多态具有更好的灵活性。比如有的动物只会叫不会飞，比如狗、猫、绵羊；有的动物只会飞不会叫，比如蝴蝶；而有的动物既会飞也会叫，比如喜鹊。这就可以通过定义两个接口 Speakable 和 Flyable，通过二者组合再声明实现类。首先声明三个实现类 Dog、Cat 和 Sheep 实现 Speakable，这三个类定义了能叫的动物。然后声明蝴蝶类 Butterfly 实现 Flyable，最后让喜鹊类 Magpie 既实现 Speakable 接口，又实现 Flyable 接口。

首先定义接口 Speakable 和 Flyable，接口 Speakable 中声明了抽象方法 speak()，接口 Flyable 中声明了抽象方法 fly()，如源文件 4-29 和源文件 4-30 所示。

源文件 4-29　Speakable.java

```
1 public interface Speakable {
2     void speak();
3 }
```

源文件 4-30　Flyable.java

```
1 public interface Flyable {
2     void fly();
3 }
```

然后声明三个实现类 Dog、Cat 和 Sheep，如源文件 4-31 至源文件 4-33 所示。这些类中定义了对抽象方法 speak() 的实现。

源文件 4-31　Dog.java

```
1 public class Dog implements Speakable {
2     public void speak() {
3         System.out.println("汪汪");
4     }
5 }
```

源文件 4-32　Cat.java

```
1 public class Cat implements Speakable {
2     public void speak() {
3         System.out.println("喵喵");
4     }
5 }
```

源文件 4-33　Sheep.java

```
1 public class Sheep implements Speakable {
2     public void speak() {
3         System.out.println("咩咩");
4     }
5 }
```

然后定义蝴蝶（Butterfly）作为接口 Flyable 的实现类，如源文件 4-34 所示。

源文件 4-34　Flyable.java

```
1 public class Butterfly implements Flyable {
2     public void fly() {
3         System.out.println("蝴蝶翩翩飞");
4     }
5 }
```

接着定义喜鹊（Magpie）类既实现 Flyable 接口，也实现 Speakable 接口，如源文件 4-35 所示。

源文件 4-35　Magpie.java

```
1 public class Magpie implements Flyable, Speakable {
2     public void speak() {
3         System.out.println("喜鹊叫喳喳");
4     }
5
6     public void fly() {
7         System.out.println("喜鹊飞呀飞");
8     }
9 }
```

客户程序创建各类动物，然后让这些动物能叫的叫，能飞的飞，如源文件 4-36 所示。

源文件 4-36　AnimalsDemo.java

```
1  public class AnimalsDemo {
2      public static void main(String[] args) {
3          Speakable a, b, c, d;
4          Flyable f, e;
5          a = new Dog();
6          b = new Cat();
7          c = new Sheep();
8          d = new Magpie();
9          e = new Butterfly();
10         a.speak();
11         b.speak();
12         c.speak();
13         d.speak();
14
15         f = (Flyable) d;
16         f.fly();
17         e.fly();
18     }
19 }
```

在源文件 4-36 中首先声明了 Speakable 类型的 4 个引用类型的变量 a、b、c、d 准备引用会叫的动物，然后依次创建了 Dog、Cat、Sheep 和 Magpie 类型的对象，并分别赋值给变量 a、b、c、d。

当在第 10 行至第 13 行分别调用 a、b、c、d 所引用的对象上的 speak() 方法时，由于运行时刻这三个对象的类型分别是 Dog、Cat 和 Sheep，所以分别执行了这些类型定义的 speak() 方法。因此输出结果各不相同。

在第 15 行把 Magpie 类型的对象转型为 Flyable，使用 Flyable 类型的变量 f 引用 Magpie 类型的对象，分别通过 Flyable 类型的引用变量调用 Butterfly 对象上的 fly() 实例方法。输出结果为：

汪汪
喵喵
咩咩
喜鹊叫喳喳
喜鹊飞呀飞
蝴蝶翩翩飞

喜鹊类 Magpie 通过实现多个接口，实现了既能叫、又能飞的设计。所以，接口与实现类的分离是提高程序可扩展性的重要技术。

4.9　源代码的组织和访问控制

Java 应用程序是一些类的集合。当类的数量很大的时候，比如成百上千个，就需要分门别类把一组相关的类型（类、接口等）定义在一起形成一个集体，称为包（package）。这样，一个类的全名（fully qualified name）就形如：<包名>.<类名>。

类名的前面有了包名的限制，就为能够避免类名冲突。当两个类名字都是 Car 时，只需把这两个类声明在不同的包里即可。

但是，如果类名相同，包名也相同，名字冲突依然存在。为了解决名字冲突的问题，需要包的名字具有唯一性，最好是在整个世界里是唯一的。如何起一个"唯一"的名字呢？从互联网主机的域名得到启发，约定以反转互联网域名作为包的名字。例如某 IT 公司 abc 的互联网域名可能是 abc.com.cn，那么该公司研发的包就应命名为 cn.com.abc。注意，包名一律使用小写字母。包名仅仅是一个标识符，其中的"."也是普通字符，没有其他含义。

JDK 包含了丰富的预先写好的类，这些类称为应用程序接口，简称 API（Application Programming Interfaces）。JDK 中的 API，也称为标准 API，为程序员编写网络、图形等应用提供了通用的类，组织在不同的以 java 为前缀的包中。例如 java.lang 包是编写 Java 程序最基本的包，其中包含了 String、Math、System 等最常用的类。java.util 包含了 Calendar、Date、Scanner、Stack、Vector、Set 和 Queue 等实用类。java.io 包中有完成字节方式输入输出和字符方式输入输出的类。例如 InputStream、OutputStream、Reader、Writer。

当程序员开始设计类的时候，就应当规划好这个类声明在哪个包中。这些由程序员设计的包称为自定义包。使用保留字 package 定义包。例如：

```
package cn.com.abc;
public class A {
    ……
}
```

源文件中的包定义语句约束了本源文件中定义的所有类和接口都属于这个包。本例中类 A 就属于包 cn.com.abc。

一个源文件中最多有一个包定义语句。如果在源文件中有一个包定义语句，则该语句必须是第 1 条语句。如果在源文件中没有包定义语句，则把类声明在默认的包中。

Java 中的包与操作系统中的文件夹有映射关系。例如包 cn.com.abc 与文件夹 cn\com\abc 相对应。包 cn.com.abc 中的类型的源文件都保存在文件夹 cn\com\abc 中。

把类定义在包中后，就有两种方式使用包中的类：全名方式和导入（import）方式。

假设包 cn.com.xyz 中的类 E 偶尔用一次包 cn.com.abc 中的类 A，就可以使用类的全名 cn.com.abc.A：

```
package cn.com.xyz;
class E {
    void accessOtherClassInOtherPackage() {
        cn.com.abc.A  a = new cn.com.abc.A();
        ……
    }
}
```

当偶尔使用其他包中的类，使用完整类名（fully qualified name）没什么问题。但当频繁使用其他包中的类时，使用完整类名就显得冗长乏味，程序读起来变得吃力。为了解决这个问题，只需在导入时使用完整类名一次，以后就使用类名而不需要包名作为前缀。导入由 import 保留字完成：

```
package cn.com.xyz;
import cn.com.abc.A;
class E {
    void accessOtherClassInOtherPackage() {
        A a = new A();
    }
}
```

每个 Java 源文件默认导入包 java.lang 中的类,无须使用 import 语句导入。使用其他 JDK 中的类时需要先导入。例如使用 java.util 包中的 Scanner 类:

```
import java.util.Scanner;
public class A {
    Scanner sc = new Scanner(System.in);
    ……
}
```

importcn.com.abc.* 导入 cn.com.abc 包中的所有类,但这不是好的程序设计实践。建议逐一导入所需要的类。import 语句告诉编译器到哪个包里去寻找某个类的定义,如果该类在当前包中找不到的话。注意,import 语句必须出现在包定义和类定义之间。

一个类要么是公共的、要么是包私有的。如果类声明为 public,这意味着该类可以被任意类使用;如果在定义类的时候没有给出访问修饰符 public,就意味着该类是包私有的(package-private),仅仅能够被同一包中的类使用。

类中的成员可以是公共的、保护的、包私有的或类私有的。公共、保护和类私有的成员分别由保留字 public、protected 和 private 修饰。公共的成员允许被任意类访问;而类私有的成员只允许在类内访问。如果类成员前没有任何访问修饰符,即默认为包私有,可被同一包中的类访问。访问性修饰符为 protected 的类成员不仅能够被同一类中的成员方法访问,还能被其所属类的子类访问,也能被同一包中类的成员方法访问。

源文件 4-37 把访问修饰符放在成员方法上,以同一包、不同类对成员方法的访问展示了可访问性的不同。

源文件 4-37　AccessModifierDemo.java

```
1  class A {
2      public void publicMember() {
3          System.out.println("I am a public member in A.");
4      }
5
6      protected void protectedMember() {
7          System.out.println("I am a protected member in A.");
8      }
9
10     void defaultMember() {
11         System.out.println("I am a default member in A.");
12     }
13
14     private void privateMember() {
15         System.out.println("I am a private member in A.");
16     }
17
```

```
18    void accessInClass() {
19        publicMember();
20        protectedMember();
21        defaultMember();
22        privateMember();
23    }
24 }
25
26 class B {
27     void accessAbyOtherClass() {
28         A a = new A();
29         a.publicMember();
30         a.protectedMember();
31         a.defaultMember();
32
33         //错误：privateMember()类 A 私有
34         a.privateMember();
35     }
36 }
37
38 class C extends A {
39     void accessAbySubClass() {
40         publicMember();
41         protectedMember();
42         defaultMember();
43         privateMember(); //错误：找不到符号
44     }
45 }
46
47 public class AccessModifierDemo {
48     public static void main(String[] x) {
49         A a = new A();
50         a.accessInClass();
51         B b = new B();
52         b.accessAbyOtherClass();
53         C c = new C();
54         c.accessAbySubClass();
55     }
56 }
```

在源文件 4-37 中，类 A 声明了成员方法分别是公共的、保护的、包私有的和类私有的。这 4 个成员方法都能被类 A 中的方法 accessInClass 访问。

类 B 的成员方法 accessAbyOtherClass()通过实例化类 A 的对象，访问其上的 4 个方法，其中访问私有成员方法会导致编译错误：privateMember()类 A 私有。

类 A 的子类 C 的成员方法 accessAbySubClass()直接调用了继承来的类 A 的成员方法。因为父类的私有成员不能被继承，所以访问类 A 的私有方法会导致编译错误：找不到符号。

源文件 4-38、源文件 4-39 和源文件 4-40 展示了不同包中的可访问性。

源文件 4-38　A.java

```
1 package cn.com.abc;
2 public class A {
3     public void publicMember() {
4         System.out.println("I am a public member in A.");
5     }
6
7     protected void protectedMember() {
8         System.out.println("I am a protected member in A.");
9     }
10
11    void defaultMember() {
12        System.out.println("I am a default member in A.");
13    }
14
15    private void privateMember() {
16        System.out.println("I am a private member in A.");
17    }
18
19    void accessAinClass() {
20        publicMember();
21        protectedMember();
22        defaultMember();
23        privateMember();
24    }
25 }
```

源文件 4-39　B.java

```
1 package cn.com.xyz;
2 import cn.com.abc.A;
3
4 public class B {
5     void accessAbyOtherClass() {
6         A a = new A();
7         a.publicMember();
8
9         //错误：protectedMember()在类A中的访问控制修饰符是protected
10        a.protectedMember();
11
12        //错误：defaultMember()在A中的访问控制修饰符不是public;不能在其他包访问
13        a.defaultMember();
14
15        //错误：privateMember()在A中的访问控制修饰符是private
16        a.privateMember();
17    }
18 }
```

源文件 4-40　C.java

```
1 package cn.com.xyz;
2 import cn.com.abc.A;
3
4 public class C extends A {
```

```
5   void accessAbySubClass() {
6       publicMember();
7       protectedMember();
8
9       //错误：找不到符号
10      defaultMember();
11
12      //错误：找不到符号
13      privateMember();
14  }
15 }
```

在源文件 4-38 中，类 A 定义在包 cn.com.abc 中，类 A 的 4 个成员方法分别是公共的、保护的、包私有的和类私有的。在源文件 4-39 中，类 B 定义在包 cn.com.xyz 里。在类 B 的 accessAbyOtherClass()方法中试图访问类 A 的实例方法，但仅能访问 public 修饰的实例方法，访问其他实例方法则出现编译错误(见第 10、13、16 行调用语句上方的注释)。

在源文件 4-40 里的子类 C 中的 accessAbySubClass()方法中试图访问继承的成员，但仅能访问 public 和 protected 修饰的实例方法。访问其他实例方法则出现编译错误(第 9 行的注释和第 12 行的注释)。

继承来的成员的可访问性只能被提升，而不能下降。在源文件 4-41 中，类 P 里声明了 public 的成员方法 f()，而在第 13 行其子类 C 试图覆盖 f()并把访问修饰符改成 private，这会导致编译错误：试图降低访问权限(attempting to assign weaker access privileges)。

源文件 4-41 C.java

```
1 class P {
2     P() {
3         System.out.println("I am a P.");
4     }
5
6     public int f(int i) {
7         return i * i;
8     }
9 }
10
11 public class C extends P {
12     @Override
13     private int f(int i) {
14         return i + i;
15     }
16
17     public static void main(String argv[]) {
18         C c = new C();
19         c.f(3);
20     }
21 }
```

为了满足面向对象程序设计的"封装性"原则，通常将类的成员变量设计为私有的，不能直接访问，但可以通过 public 的 getter 和 setter 方法间接访问。比如类 Car 把成员变量 licenceNumber 设计为 private，为了能够让其他类访问这个成员，提供了 getter 方法：

```
public String getLicenceNumber() {
    return licenceNumber;
}
```

还提供了 setter 方法:

```
public void setLicenceNumber(String licenceNumber) {
    this.licenceNumber = licenceNumber;
}
```

getter 方法的方法名由 get 和成员变量名两部分拼接而成,成员变量名首字母大写; setter 方法的方法名由 set 和成员变量名两部分拼接而成,成员变量名首字母大写。当成员变量的类型是 boolean 时,情况有些特殊:getter 方法的方法名由 is 和成员变量名组成。例如,下面的类 A 中有布尔类型的成员变量 active,其 getter 方法的名字为 isActive。

```
public class A {
    boolean active;

    public boolean isActive() {
        return active;
    }

    public void setActive(boolean active) {
        this.active = active;
    }

}
```

4.10 Object 类

java.lang.Object 类是所有类的祖先类。进行类的定义时,即便没有显式地继承 Object 类,Java 也默认继承了 Object。即

```
public class Car {
}
```

等价于:

```
public class Car extends Object {
}
```

所以,当定义了一个类后,这个类默认继承了 Object 类的一些方法,包括 toString()、clone()、equals(Object obj)、hashCode()等。

4.10.1 toString 方法

toString 方法返回对象的字符串表示。Object 提供的对象的默认字符串表示形如"<类名>@<对象哈希码>"。其中,<类名>是实例化该对象时所使用的类;而默认的对象哈希码就是对象所在的内存地址,以十六进制表示。例如源文件 4-42 中的 main 方法首先实例化

了类 A 的对象,然后输出这个对象的字符串表示。

源文件 4-42　A.c

```
1  public class A {
2      public static void main(String[] args) {
3          A a = new A();
4          System.out.println(a);
5      }
6  }
```

System.out.println(a)等价于 System.out.println(a.toString())。即 println 方法会自动调用引用类型变量所引用对象的 toString 方法。此例的一个可能的输出是:

A@2f92e0f4

一个实体类需要默认覆盖继承的 Object 的 toString 方法以方便使用 System.out.println()输出对象的字符串表示。在源文件 4-43 中的类 Car 中声明了 4 个成员变量,其 toString 方法以"<变量名>=<变量值>"的格式显示每个成员变量的名字及其值,并以逗号隔开。所有的名值对使用方括号括起来,并使用类的名字作为前缀。一个可能的输出结果为:

Car [licenceNumber =1234, length=4500, width=2300, height = 1600]

源文件 4-43　Car.java

```
1  public class Car {
2      private String licenceNumber;
3      private int length, width, height;
4
5      Car(String licenceNumber, int length, int width, int height) {
6          this.licenceNumber = licenceNumber;
7          this.length = length;
8          this.width = width;
9          this.height = height;
10     }
11
12     public String toString() {
13         return "Car [licenceNumber =" + licenceNumber
14             + ", length=" + length
15             + ", width=" + width
16             + ", height = " + height
17             + "]";
18     }
19
20     public static void main(String[] args) {
21         Car c = new Car("1234", 4500, 2300, 1600);
22         System.out.println(c);
23     }
24 }
```

4.10.2　equals 方法

Java 中约定使用 equals 方法用于判断两个对象是否相等。对于基本数据类型的变量,两个变量相等指的是这两个变量中存储的值相等;而对象通过引用类型变量来引用。比如

下面的代码声明了两个引用类型变量 a 和 b, 分别引用两个类 A 的对象:

```
A  a, b;
a = new A();
b = new A();
```

如果判断变量 a 引用的对象和变量 b 引用的对象是否相等, 则使用对象上的 equals 方法:

```
a.equals(b)
```

在深入介绍 equals 方法的用法之前, 先介绍两个概念: 同一(identical)和相等(equal)。"同一"就是指两个引用类型变量 a 和 b 是否引用了同一个对象。比如下面的代码片段就使得引用类型变量 a 和 b 引用了同一个对象:

```
A  a, b;
a = new A();
b = a;
```

此时的内存布局如图 4-5 所示。

图 4-5　同一

判断两个引用类型变量是否引用同一个对象, 只需使用关系运算符"=="即可:

```
a == b
```

如果引用类型变量 a 和 b 引用了同一个对象, 那么 a 和 b 引用的对象相等。"同一"对象当然"相等"。

当引用类型变量 a 和 b 分别引用各自的对象, 如图 4-6 所示, 这两个对象是否相等则取决于类 A 如何覆盖 Object 类的 equals 方法。

图 4-6　相等

Object 类提供的 equals 方法仅仅进行"同一"判断:

```
public boolean equals(Object obj){
    return (this == obj)
}
```

也就是说, 默认从 Object 类继承的 equals 方法的语义与"=="运算符相同。但是大多数情形同一对象仅仅是相等对象的特殊情形。一般来说如果两个对象具有相同的状态, 则认为这两个对象相等。具有相同的状态并不意味着两个对象所有实例变量的取值必须相等。比如两个 Car 类的对象，可以认为车牌号相等，那么这两个对象就相等；也可以认为如果质量（weight）相等，这两个对象就相等。所以具体如何定义对象相等，取决于程序所解决的实际问题。下面的类 Car 的定义中就覆盖了 Object 类的 equals 方法：

```
class Car {
    private int weight;
    private String licenceNumber;
    public boolean equals(Object obj) {
        if (this == obj)
            return true;
        if (obj == null)
            return false;
        if (getClass() != obj.getClass())
            return false;

        //确保此刻 obj 引用的是 Car 类型的对象
        Car other = (Car) obj;
        if (licenceNumber == null) {
            if (other.licenceNumber != null)
                return false;
        } else if (! licenceNumber.equals(other.licenceNumber))
            return false;
        if (weight != other.weight)
            return false;
        return true;
    }
}
```

覆盖 equals 方法比较对象 a 和 b，即 a.equanls(b)，一般通过如下步骤完成。

（1）首先使用关系运算符"=="比较对象 a 和 b 的同一性，如果同一，则返回真。

（2）如果 b 为 null 则返回假。

（3）如果 b 的类型与 a 的类型不同，则返回假。获取对象的类型通过从 Object 类继承的方法 getClass 完成。

（4）把 b 转型为 a 的类型。上例中把 Object 类型转型为 Car 类型。

（5）逐一比较选择的实例变量。当选择的实例变量都相等，则返回真。基本数据类型的实例变量直接使用比较运算符"=="；引用类型的实例变量仍需调用其上的 equals 方法。此例中在调用字符串对象上的 equals 方法之前检查了是否为空，这是为了避免出现 NullPointerException 异常。

注意：equals 方法的形式参数类型是 java.lang.Object。一定不要使用其他类型，否则就无法覆盖 java.lang.Object。

equals 方法其实定义了一个等价关系。等价关系是自反的（reflexive）、对称的（symmetric）和传递的（transitive）。自反的意思是自己与自己相等。对称的意思是如果对象 a 与 b 相等，那么 b 与 a 相等。传递的意思是如果对象 a 与 b 相等，b 与 c 相等，那么 a 与 c 相等。

JDK 中很多类都提供了 equals 方法。例如源文件 4-44 是对当地日期 LocalDate 类的对象进行比较。

源文件 4-44　EqualsDemo.java

```
1 import java.time.LocalDate;
2 public class EqualsDemo {
3     public static void main(String[] args) {
4         LocalDate a = LocalDate.of(2024,1,22);
5         LocalDate b = LocalDate.of(2024,1,22);
6         System.out.println(a.equals(b));
7         System.out.println(a == b);
8     }
9 }
```

在这个例子中，首先实例化了两个 2024 年 1 月 22 日的 LocalDate 对象，分别使用 a 和 b 引用。此时两个 LocalDate 类型对象的年、月、日分别相等，equals 方法返回真。但是 a 和 b 引用的是两个不同的对象，所以程序的输出是：

```
true
false
```

前一个 true 是第 6 行的输出；后一个 false 是第 7 行的输出。类 LocalDate 中声明的 equals 方法调用 compareTo 方法来比较两个 LocalDate 对象的年、月、日是否都相等：

```
int compareTo(LocalDate otherDate) {
    int cmp = (year - otherDate.year);
    if (cmp == 0) {
        cmp = (month - otherDate.month);
        if (cmp == 0) {
            cmp = (day - otherDate.day);
        }
    }
    return cmp;
}
```

表 4-1 比较了对象同一的情形和对象相等的情形。

表 4-1　对象的同一比较和相等比较

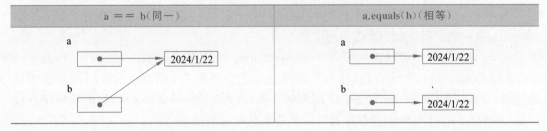

4.10.3　hashCode 方法

hashCode 方法主要用于哈希表等存储结构。当把对象添加到这些集体对象中去的时候会调用对象的 hashCode() 方法来计算在哈希表中的地址。Object 类中提供了 hashCode 的默认实现：以十六进制返回对象的内存地址。

hashCode() 方法应遵循以下规则。

(1) 相等对象必须产生相同的哈希码（一致性）：如果 a.equals(b) 为 true，那么 a.hashCode() 和 b.hashCode() 应该返回相同的整数值。

(2) 不同对象尽量产生不同的哈希码（扩散性）：虽然无法保证完全避免哈希冲突，但一个好的哈希函数应该让不同对象的哈希码分布均匀，以减少冲突的概率。

(3) 哈希码应该是稳定的：如果对象内容不变，则其哈希码不应改变。即只要对象状态不变，调用其 hashCode 方法的返回值就不变，无论调用几次。

在自定义类中，如果覆盖了 equals() 方法，那么也应该同时覆盖 hashCode() 方法，以确保两个方法之间的逻辑一致。

一种计算哈希码的算法是使用一个素数迭代地与所有实例变量累加和相乘。例如：

```
class Car {
    private int length, width;
    public int hashCode() {
        final int prime = 31;
        int result = 1;
        result = prime * result + length;
        result = prime * result + width;
        return result;
    }
}
```

这个例子首先设置素数 31，初始的累加和为 1。然后把累加和乘以 31，再与 length 累加；再把累加和乘以 31，再与第二个实例变量 width 累加。

java.lang.String 类提供的 hashCode 方法把字符串看作是"三十一进制数"，其哈希码是

$$s_0 \times 31^{n-1} + s_1 \times 31^{n-2} + \cdots + s_{n-1}$$

其中，s_i 是字符串中的第 i 个字符，n 是字符串的长度。空字符串的哈希码是 0。

4.11 枚举

枚举（enum）是一种特殊的类（class），用于定义一组相关的离散量。enum 是 enumerations 的简写，意思是"逐一列出"。

enum 定义的枚举类默认继承了 java.lang.Enum 类，并实现了 java.lang.Serializable 和 java.lang.Comparable 两个接口。java.lang.Enum 类中声明了实例方法 ordinal 返回枚举量的索引。编译器会在 enum 类型中插入一些方法，包括类方法 values() 以数组返回所有枚举量，类方法 valueOf(String) 根据枚举名字符串获取枚举量等。

源文件 4-45 中声明了一个名为 Color 的枚举类型，逐一列出了三个枚举量：红、黄、绿，用来作为交通信号灯颜色的类型。注意，枚举量其实就是 Color 类型的对象：

```
class Color {
    public static final Color RED = new Color();
    public static final Color YELLOW = new Color();
    public static final Color GREEN = new Color();
}
```

这些对象通过引用变量 RED、YELLOW 和 GREEN 进行引用。由于引用变量由 public static final 修饰，所以要大写。

第 7 行输出 RED 所引用的枚举量的字符串表示，这会调用父类 Enum 的实例方法 toString。

第 8 行调用静态方法 values 把所有的枚举量放到数组中返回，然后使用 for-each 循环遍历输出。

第 13 行的 switch 语句试图输出枚举量的中文标签。例如 RED 的中文标签是"红色"。

源文件 4-45　TestColor.java

```
1 enum Color {
2     RED, YELLOW, GREEN;
3 }
4
5 public class TestColor {
6     public static void main(String[] args) {
7         System.out.println(Color.RED);
8         for (Color color : Color.values()) {
9             System.out.println(color);
10        }
11
12        Color color = Color.RED;
13        switch (color) {
14            case RED -> System.out.println("红色");
15            case YELLOW -> System.out.println("黄色");
16            case GREEN -> System.out.println("绿色");
17        }
18    }
19 }
```

enum 声明实际上定义了一个类。因此可以在类中声明其他成员。比如在枚举类 Color 中声明各个枚举量的中文标签，如源文件 4-46 所示。把枚举量的名字与中文标签的映射定义在枚举类型中提高了程序的复用程度和可读性。

第 3 行逐一声明各个枚举量的引用变量名时还给出了中文标签作为参数，并在第 9 行定义了构造方法，这个构造方法就把字符串类型的中文标签作为参数创建枚举量，中文标签存储在实例变量 label 中，即每一个枚举量都有自己的成员变量 label。

第 13 行声明实例方法 getLabel 返回中文标签。

第 18 行声明实例方法 getByName 根据枚举量的字符串名返回枚举量，比如根据字符串 YELLOW 返回枚举量 Color.YELLOW。第 21 行的实例方法 name 是从 Enum 继承的方法，方法以字符串返回枚举量的名字。

第 32 行直接让变量 trafficLight 引用枚举量 Color.RED。第 33 行输出该枚举量的字符串表示。第 34 行调用实例方法 getLabel 返回该枚举量的中文标签。第 35 行根据枚举量的字符串名 YELLOW 查找枚举量 Color.YELLOW。

源文件 4-46　EnumDemo.java

```
1 enum Color {
2     //逐一列出各个枚举量
3     RED("红色"), YELLOW("黄色"), GREEN("绿色");
```

```
4
5       //枚举量的标签
6       private String label;
7
8       //构造方法,构造枚举量对象
9       Color(String label) {
10          this.label = label;
11      }
12
13       public String getLabel() {
14          return label;
15      }
16
17      //根据枚举量的字符串名查找枚举对象
18      public static Color getByName(String name) {
19          Color result = null;
20          for (Color color : values()) {
21              if (color.name().equalsIgnoreCase(name)) {
22                  result = color;
23                  break;
24              }
25          }
26          return result;
27      }
28 }
29
30 public class EnumDemo {
31     public static void main(String[] args) {
32          Color trafficLight = Color.RED;
33          System.out.println(trafficLight);
34          System.out.println(trafficLight.getLabel());
35          System.out.println(Color.getByName("YELLOW"));
36      }
37 }
```

程序的输出为:

```
RED
红色
YELLOW
```

第 4 章 章节测验

第 5 章 异 常

正常(normal)的程序沿着控制流执行。在执行过程中可能会发生某事件导致程序无法继续沿着控制流执行,这种不期望发生的事件称为异常(exception)。异常是"异常事件"(exceptional event)的简称。比如使用索引 3 访问长度为 3 的数组就会抛出"数组越界"异常。

5.1 异常的抛出与捕获

程序把异常事件创建为"异常对象"并传递给 Java 虚拟机,这个过程称为"抛出"(throw)。异常对象中包含了错误信息以及程序在异常出现时的状态。异常是"少而不同"的事件,并不总会发生。如果程序需要对可能发生的异常事件按预案进行处理,那么程序就设计一段代码首先从虚拟机中取出异常对象,这个过程称为"捕获"(catch),然后再进行处理。图 5-1 展示了异常从正常流程语句中抛出到虚拟机,用于处理异常的程序块从虚拟机中捕获异常再处理。从示意图可以看出,Java 的异常机制把程序的正常流程和异常处理流程分离,提高了源代码的可读性。

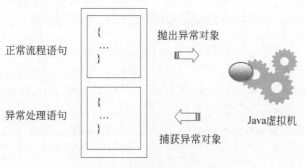

图 5-1 异常的抛出与捕获

数组越界异常 ArrayIndexOutOfBoundsException 是在 java.lang 包中定义的类。源文件 5-1 演示了数组越界异常的产生。在第 3 行首先创建了一个只具有三个元素的数组对象，即数组的长度为 3，按照 Java 语言对数组的访问规则，应分别通过 a[0]、a[1] 和 a[2] 访问这三个数组对象。然而第 4 行的语句却试图访问 a[3]。当程序执行到第 4 行的语句时，println 方法就会向虚拟机抛出 ArrayIndexOutOfBoundsException 类型的异常对象，并在控制台输出运行时刻错误消息：

```
Exception in thread "main" java.lang.ArrayIndexOutOfBoundsException: Index 3
out of bounds for length 3
at ArrayIndexOutOfBoundsExceptionDemo.main
(ArrayIndexOutOfBoundsExceptionDemo.java:4)
```

意思是在源文件 ArrayIndexOutOfBoundsExceptionDemo.java 第 4 行 main 方法中出现 java.lang.ArrayIndexOutOfBoundsException 类型的异常：索引 3 超出了长度为 3 的数组的索引范围。

由于程序没有设计任何预案来应对该异常，程序终止执行。

源文件 5-1　ArrayIndexOutOfBoundsExceptionDemo.java

```
1 public class ArrayIndexOutOfBoundsExceptionDemo {
2     public static void main(String[] args) {
3         int[] a = {10, 20, 30};
4         System.out.println(a[3]);
5     }
6 }
```

除了数组越界异常，常见异常还有 NullPointerException，如源文件 5-2 第 4 行试图访问静态变量 a；而 a 没有引用任何数组对象，其初值为 null。所以当程序运行到第 4 行会抛出空指针异常，并在控制台输出如下消息后终止执行：

```
Exception in thread "main" java.lang.NullPointerException: Cannot load from int
array because "NullPointerExceptionDemo.a" is null
    at NullPointerExceptionDemo.main(NullPointerExceptionDemo.java:4)
```

这段消息说，在源文件 NullPointerExceptionDemo.java 第 4 行出现 java.lang.NullPointerException 类型的异常：因为 a 的值是 null，所以无法从整型数组读取。

源文件 5-2　NullPointerExceptionDemo.java

```
1 public class NullPointerExceptionDemo {
2     static int[] a;
3     public static void main(String[] args) {
4         System.out.println(a[0]);
5     }
6 }
```

java.lang.NullPointerException 的父类是运行时刻异常 java.lang.RuntimeException；java.lang.RuntimeException 的父类是 java.lang.Exception；java.lang.Exception 的父类是 java.lang.Throwable。RuntimeException 是 ArrayIndexOutOfBoundsException 的祖先类。JDK 中的异常继承关系如图 5-2 所示。

从图 5-2 中看到，异常是可抛出的（Throwable）对象。运行时刻异常 RuntimeException 也

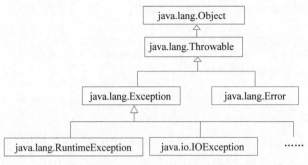

图 5-2　JDK 中的异常

称为免检（unchecked）异常，而 Exception 类的其他子类都是受检（checked）异常。所谓"免检"，指这类异常导致程序无法继续执行下去，只能终止，不必为这类异常设计处理程序；所谓"受检"，指程序必须安排预案，一旦异常出现如何处置。IOException 是典型的受检异常。FileNotFoundException 是一种 IOException。源文件 5-3 演示了编译器如何强制程序处理可能出现的 IOException 对象。

源文件 5-3　FileNotFoundExceptionDemo.java

```
1 import java.util.Scanner;
2 import java.io.File;
3 public class FileNotFoundExceptionDemo {
4     public static void main(String[] args) {
5         Scanner sc = new Scanner(new File("input.txt"));
6         System.out.println(sc.nextLine());
7         sc.close();
8     }
9 }
```

源文件 5-3 试图从文本文件 input.txt 中读取一行。第 5 行从当前路径中读取文本文件 input.txt，所以以文件名 input.txt 为实参创建 File 类型的对象以准备访问磁盘文件 input.txt，然后把 File 对象作为实参创建 Scanner 对象以从该文件中读写数据。第 5 行会出现编译错误：

```
FileNotFoundExceptionDemo.java:5: error:
unreported exception FileNotFoundException; must be caught or declared to be thrown
```

意思是在源文件 FileNotFoundExceptionDemo.java 第 5 行有错误：找不到文件异常 FileNotFoundException 要么被捕获、要么声明被抛出。

这样，编译器强制在程序中必须对可能出现的 IOException 进行处理。处理的方法有两种：捕获和声明抛出。声明抛出是较为简单的处理方法：如果在方法体中可能抛出某类型的异常，那么只需把该异常类型在方法头部使用保留字 throws 声明抛出即可。该方法的调用者看到这个声明，再进行处理。如源文件 5-4 第 5 行所示。

源文件 5-4　FileNotFoundExceptionDemo.java

```
1 import java.util.Scanner;
2 import java.io.File;
3 import java.io.FileNotFoundException;
```

```
4 public class FileNotFoundExceptionDemo {
5     public static void main(String[] args)
          throws FileNotFoundException {
6         Scanner sc = new Scanner(new File("input.txt"));
7         System.out.println(sc.nextLine());
8         sc.close();
9     }
10 }
```

File 和 FileNotFoundException 都是 java.io 包中的类,注意使用 import 保留字导入。

另外一种处理方式就是使用保留字 catch 捕获异常,然后在 catch 后面的块中处理该异常。源文件 5-5 使用 try-catch 捕获和处理受检异常 FileNotFoundException。

源文件 5-5 CatchDemo.java

```
1 import java.util.Scanner;
2 import java.io.File;
3 import java.io.FileNotFoundException;
4 public class CatchDemo {
5     public static void main(String[] args) {
6         Scanner sc = null;
7         try {
8             sc = new Scanner(new File("input.txt"));
9             System.out.println(sc.nextLine());
10            sc.close();
11        } catch (FileNotFoundException e) {
12            e.printStackTrace();
13            System.out.println(e.getMessage());
14        }
15    }
16 }
```

把可能抛出异常的程序片段放在保留字 try 后的一对花括号"{}"中。这对花括号括起来的语句序列称为 try 块(try block)。然后把对异常处理的程序片段放在保留字 catch 后面的花括号中,这对花括号"{}"括起来的语句序列称为 catch 块。一旦 try 块中的某条语句抛出异常对象,就不再继续执行该语句以及其后随语句,而是转移到 catch 块执行。

为了在 catch 块中引用 try 块中抛出的异常对象,在保留字 catch 其后的 catch 块之间使用圆括号"()"括起来了异常对象引用变量声明。执行 catch 块时,通过声明的异常对象引用变量(此例中是 e)可以访问异常对象的具体内容。第 12 行调用了实例方法 printStackTrace 输出异常对象的栈跟踪信息,这是给程序员看的,用来调试程序的;第 13 行调用了实例方法 getMessage 输出异常提示消息,这是给最终用户看的。执行 catch 块完毕,继续执行 try-catch 语句的后随语句。程序的输出结果为:

```
java.io.FileNotFoundException: input.txt (系统找不到指定的文件。)
    at java.base/java.io.FileInputStream.open0(Native Method)
    at java.base/java.io.FileInputStream.open(FileInputStream.java:219)
    at java.base/java.io.FileInputStream.<init>(FileInputStream.java:158)
    at java.base/java.util.Scanner.<init>(Scanner.java:645)
    at CatchDemo.main(CatchDemo.java:8)
input.txt (系统找不到指定的文件。)
```

最后一行是实例方法 getMessage 输出。图 5-3 展示 try 块、catch 块、异常对象和 JVM 之间的关系。

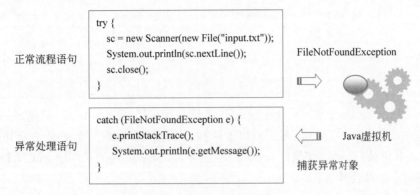

图 5-3　try 块和 catch 块

在某次运行中,try 块中的某条语句可能会抛出异常对象,此例中是 FileNotFoundException 类型的异常对象;这个异常对象交给了 JVM;程序的控制发生转移,转到 catch 块执行;catch 块通过声明的 FileNotFoundException 类型的引用变量 e 访问 try 块中抛出的异常对象,一旦 e 能够引用这个异常对象,则称捕捉到了该异常。

catch 块只能捕获 Exception 类及其子类的对象。但是 RuntimeException 及其后代类是免检异常,如 ArithmeticException、ClassCastException、IllegalArgumentException、IndexOutOfBoundsException、NegativeArraySizeException、NoSuchElementException、NullPointerException 等。除了 RuntimeException 以外,其他 Exception 类的后代类都是受检异常。对于所有受检异常,程序必须显式地声明如何进行处理。

5.2　处理异常

源文件 5-5 中的 try-catch 语句在没有任何异常出现的时候能够正确运行,但是如果 try 块中第 8 行的语句抛出异常,那么就不会执行第 10 行的关闭语句 close(),从而无法释放资源。合理的做法是将所有清理现场和释放资源的语句都放到 finally 块中或者使用 try-with-resource 语句。源文件 5-6 演示了 finally 块的用法。

源文件 5-6　**FinallyDemo.java**

```
1  import java.util.Scanner;
2  import java.io.File;
3  import java.io.FileNotFoundException;
4
5  public class FinallyDemo {
6      public static void main(String[] args) {
7          Scanner sc = null;
8          try {
9              sc = new Scanner(new File("input.txt"));
10             System.out.println(sc.nextLine());
11         } catch (FileNotFoundException e) {
```

```
12              e.printStackTrace();
13              System.out.println(e.getMessage());
14      } finally {
15              if (sc != null) {
16                  sc.close();
17              }
18          }
19      }
20 }
```

在源文件 5-6 中，如果在 try 块中没有异常抛出，那么执行 finally 块，关闭 Scanner 对象 sc；如果在 try 块中抛出了异常，那么不再继续执行 try 块中抛出异常的语句及其后面的语句，转入 catch 块执行，再继续执行 finally 块，也会关闭 Scanner 对象 sc。所以，无论异常发生与否，finally 块都会执行。

try-catch 语句的一般形式是：

```
try{
    ……
}catch(Exception <引用变量名>){
    ……
}finally{
    ……
}
```

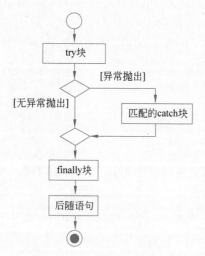

图 5-4　try-catch-finally 块

如果有 try 块，则至少与之关联一个 catch 块。另外，还可以至多与 try 块关联一个 finally 块。因为 try-catch 语句的执行有两种可能：异常不发生，仅执行 try 块；异常发生，还需执行 catch 块。当有一个 finally 与 try 块相关联时，在这两种情形下，都会执行 finally 块。也就是说，无论异常发生与否，finally 块总会执行。因此，finally 块是进行一些清除现场操作的理想位置。图 5-4 是展示 try-catch-finally 语句控制流程的 UML 活动图。

源文件 5-7 演示了 try-with-resource 的用法：把源文件 5-4 中第 6 行声明及创建 Scanner 对象的语句移到保留字 try 与花括号间的圆括号中。这样无须使用 finally 块。

源文件 5-7　WithResourceDemo.java

```
1 import java.util.Scanner;
2 import java.io.File;
3 import java.io.FileNotFoundException;
4
5 public class WithResourceDemo {
6     public static void main(String[] args) {
7         try (Scanner sc = new Scanner(new File("input.txt"));) {
8             System.out.println(sc.nextLine());
9         } catch (FileNotFoundException e) {
```

```
10          e.printStackTrace();
11          System.out.println(e.getMessage());
12      }
13   }
14 }
```

可以通过分号分隔声明的多个资源。例如源文件 5-8 试图把当前文件夹中的 input.txt 文件逐行复制到 output.txt 中。在第 8 行和第 9 行声明了 scanner 和 writer 引用两个变量分别引用 Scanner 对象和 PrintWriter 对象，这两个对象都会占用操作系统资源。

源文件 5-8　WithMultiResourceDemo.java

```
1 import java.util.Scanner;
2 import java.io.File;
3 import java.io.PrintWriter;
4 import java.io.FileNotFoundException;
5
6 public class WithMultiResourceDemo {
7     public static void main(String[] args) {
8         try (Scanner scanner = new Scanner(new File("input.txt"));
9              PrintWriter writer = new PrintWriter(new File("output.txt"))) {
10            while (scanner.hasNext()) {
11                writer.print(scanner.nextLine());
12            }
13        } catch (FileNotFoundException e) {
14            e.printStackTrace();
15            System.out.println(e.getMessage());
16        }
17    }
18 }
```

catch 块中处理异常的代码称为异常处理器(exception handler)。一个 try 块可以与一个或多个 catch 块相关联。例如源文件 5-9 试图从当前文件夹的文本文件 input.txt 中读取一行写入 output.txt 中。如果当前文件夹中没有 input.txt 则抛出 FileNotFoundException 异常；如果有 input.txt 但里面是空的，一行也没有，则抛出 NoSuchElementException 异常；如果试图读取 input.txt 时使用的对象已经关闭，则抛出 IllegalStateException 异常。程序期望捕获这三种异常就可通过多个 catch 块与 try 块关联来实现，如源文件 5-9 所示。

源文件 5-9　MultiCatchDemo.java

```
1 import java.util.Scanner;
2 import java.io.File;
3 import java.io.PrintWriter;
4 import java.io.FileNotFoundException;
5 import java.util.NoSuchElementException;
6
7 public class MultiCatchDemo {
8     public static void main(String[] args) {
9         try (Scanner scanner = new Scanner(new File("input.txt"));
10             PrintWriter writer = new PrintWriter(new File("output.txt"))) {
11            writer.print(scanner.nextLine());
12        } catch (FileNotFoundException e) {
13            e.printStackTrace();
```

```
14                System.out.println(e.getMessage());
15            } catch (NoSuchElementException e) {
16                e.printStackTrace();
17                System.out.println(e.getMessage());
18            } catch (IllegalStateException e) {
19                e.printStackTrace();
20                System.out.println(e.getMessage());
21            }
22        }
23 }
```

在源文件 5-9 中，如果 try 块中抛出异常，Java 的异常处理系统就自上而下依次将该异常的类型与 catch 中声明的类型进行匹配（match）。一旦匹配，就执行相应的 catch 块，catch 块执行完毕，则整个 try-catch 语句执行完毕，继续执行该语句的后随语句。匹配的意思是：抛出的异常对象的引用能够合法地赋值异常处理器声明的参数（e）。匹配并不要求抛出异常的类型与异常处理器声明的异常类型严格相同。如果抛出的异常是子类，而异常处理器声明的类型是父类，也认为匹配。

在源文件 5-9 中，一个 try 块与三个 catch 块相关联。自上而下，catch 块声明的异常类型分别是 FileNotFoundException、NoSuchElementException 和 IllegalStateException。如果在 try 块中抛出了 FileNotFoundException 类型的异常对象，由于该类型与 FileNotFoundException 匹配，那么执行的是第一个 catch 块；如果在 try 块中抛出了 NoSuchElementException 类型的异常对象，由于该类型与 NoSuchElementException 匹配，那么执行的是第二个 catch 块；如果在 try 块中抛出了 IllegalStateException 类型的异常对象，由于该类型与 IllegalStateException 匹配，那么执行的是第三个 catch 块。

也可以把期望捕获的异常通过"|"声明在一起，如源文件 5-10 第 12 行所示。

源文件 5-10　MultiCatchInOneDemo.java

```
1 import java.util.Scanner;
2 import java.io.File;
3 import java.io.PrintWriter;
4 import java.io.FileNotFoundException;
5 import java.util.NoSuchElementException;
6
7 public class MultiCatchInOneDemo {
8     public static void main(String[] args) {
9         try (Scanner scanner = new Scanner(new File("input.txt"));
10             PrintWriter writer = new PrintWriter(new File("output.txt"))){
11             writer.print(scanner.nextLine());
12         } catch (FileNotFoundException | NoSuchElementException
                    | IllegalStateException e) {
13             e.printStackTrace();
14             System.out.println(e.getMessage());
15         }
16     }
17 }
```

注意，由于子类的对象总能够上转型为祖先类的引用变量，下面的 try-catch 语句中所有期望的异常都能赋值给第一个 catch 块中的 e，导致其他 catch 块不会被执行：

```
try {
    ...
}
catch(Exception e) {
    System.out.println(e.getMessage());
}
catch(IOException e) {
    System.out.println(e.getMessage());
}
catch (FileNotFoundException e){
    System.out.println(e.getMessage());
}
```

如果抛出异常的语句 S 没有在任何 try 块内，那么 Java 会去包含该语句的方法的调用者中寻找相应的调用语句 R，检查 R 是否在 try 块中。如果是，则执行与 try 块相伴的 catch 块，catch 块执行完毕后，继续执行包含语句 R 的 try-catch 块的后随语句；如果在调用者中发现调用语句 R 没有在 try 块中，那么 Java 继续去语句 R 所在方法的调用者中寻找相应的调用语句。如果寻找到 main 方法也没有找到处理该异常的 try-catch 块，Java 的异常处理系统输出打印关于该异常的一些信息并终止程序。

总之，Java 异常处理系统的任务就是当程序中的异常发生后，为其寻找异常处理器。这个寻找是沿着方法的调用反向进行的，直到 main 方法。如果把一个方法看作一个节点，方法的调用是节点之间的有向边，那么 main 方法调用方法 methodA、methodA 方法又调用方法 methodB，等等，就形成了一个方法调用链。对异常处理器的寻找就是沿着方法调用链逆向进行的。如果直到 main 方法也没有找到异常处理器，那么异常处理系统向控制台输出一些消息，并终止程序执行。

如果抛出异常的语句在 try 块中，但是该异常不能与 try-catch 语句中的任何 catch 块声明的异常类型相匹配，那么该异常就沿着方法调用链向调用者方向传播。

使用异常要注意以下 6 点。

（1）一个 try 块至少有一个 catch 块关联，可与多个 catch 块关联。

（2）finally 块可有可无。

（3）必须声明如何处理受检异常。

（4）异常对象由程序中某个方法在执行期间抛出给 JVM，而 catch 块则声明要求 JVM 交递的异常。

（5）如果 JVM 存在某个异常对象，但没有找到等待该异常的任何 catch 块，则 JVM 中止程序运行。

（6）不能在 finally 块中使用 return。在 finally 块中的 return 返回后方法结束执行，不会再执行 try 块中的 return 语句。

5.3 自定义异常

前面介绍了如何捕获和处理异常，但没有解释如何抛出异常。通过保留字 throw 抛出异常。例如源文件 5-11 是 JDK 中 java.lang.Math 类的类方法 addExact，该方法把给定的两个 int 类型参数求和,如果加法运算的结果溢出,超出了 int 类型整数的范围,那么抛出 ArithmeticException 异常。

源文件 5-11　AddExact.java

```
1   class AddExact {
2       /**
3        * 返回两个整型参数的和
4        * 如果相加的结果溢出,则抛出异常
5        *
6        * @param x 第一个整数
7        * @param y 第二个整数
8        * @return 相加的结果
9        * @throws ArithmeticException 如果相加的结果整数溢出
10       * @since 1.8
11       */
12      public static int addExact(int x, int y) {
13          int r = x + y;
14          //当且仅当两个参数的符号都和结果的符号相反
15          if (((x ^ r) & (y ^ r)) < 0) {
16              throw new ArithmeticException("integer overflow");
17          }
18          return r;
19      }
20  }
```

当被加数与结果的符号相反而且加数也与结果的符号相反时发生溢出。整数的补码最高位是符号位:1 表示该整数为负数;是 0 则表示该整数为非负数。x^r 表示将 x 的每个二进制位与 r 的对应位进行异或操作,如果两个位相同则结果为 0,不同则结果为 1。y^r 同理,对 y 和 r 进行异或操作。(x^r)&(y^r)将上述两个异或结果进行按位与运算,只有当两个对应的位都是 1 时,结果的该位才是 1,否则为 0。如果被加数最高位和结果的最高位不同,那么异或的结果是 1;同理,如果加数最高位和结果的最高位不同,那么异或的结果也是 1;1 和 1 进行按位逻辑与运算,结果为 1。最高位为 1 的整数表示负数,负数小于 0,所以表达式((x^r)&(y^r))<0 为真,抛出异常。

源文件 5-12　MathDemo.java

```
1   public class MathDemo {
2       public static void main(String[] args) {
3           try {
4               System.out.println(addExact(2, 3));
5               System.out.println(addExact(2, Integer.MAX_VALUE));
6           } catch (ArithmeticException e) {
7               System.out.println(e.getMessage());
8           }
9       }
10
11      public static int addExact(int x, int y) {
12          int r = x + y;
13          //Overflow iff
14          //both arguments have the opposite sign of the result
15          if (((x ^ r) & (y ^ r)) < 0) {
16              throw new ArithmeticException("integer overflow");
```

```
17        }
18        return r;
19    }
20 }
```

第 4 行把 2 和 3 相加,结果没有溢出;而在第 5 行把在 java.lang.Integer 中定义的整数最大值 MAX_VALUE 即 0x7fffffff 与 2 相加,发生溢出,所以程序输出:

```
5
integer overflow
```

前文介绍了如何捕获和处理 JDK 中已经定义的异常,我们也可以自定义异常。自定义异常需要做三件事情:定义异常,抛出异常,捕获和处理异常。前两件事情做好之后,第三件事情就与前文讲述的异常的捕获和处理一样了。

一般通过继承 java.lang.Exception 定义异常类。如果希望自定义的异常免检(编译器不会强制要求捕获或抛出),可以继承 java.lang.RuntimeException 或其子类。通常需要设计一个以给最终用户看的异常提示消息为参数的构造方法,该方法通常通过调用父类的带有字符串参数的构造函数来完成,如源文件 5-13 所示。

源文件 5-13 MyCustomException.java

```
1 public class MyCustomException extends Exception {
2     //只有消息的构造方法
3     public MyCustomException(String message) {
4         super(message);
5     }
6
7     //带原因异常和消息的构造方法
8     public MyCustomException(String message, Throwable cause) {
9         super(message, cause);
10    }
11 }
```

可选地,还可以设计提示消息和异常原因两个参数的构造方法,如第 8 行所示。

然后在适当的方法中使用 throw 关键字抛出自定义异常。默认当参数小于 0 的时候,java.lang.Math 中定义的计算平方根静态方法 sqrt 返回 NaN。如果期望当参数小于 0 的时候抛出异常,那么可以设计一个抛出异常版本的 sqrt 方法,如源文件 5-14 所示。

源文件 5-14 Utils.java

```
1 public class Utils {
2     public static double sqrt(double x) throws MyCustomException {
3         if (x < 0) {
4             throw new MyCustomException("error: x < 0.");
5         }
6         //返回平方根,正数
7         //如果参数是 NaN 或者小于零,那么结果为 NaN
8         return Math.sqrt(x);
9     }
10 }
```

在第 3 行如果条件满足,则以提示消息 error:x<0 创建 MyCustomException 异常并抛出。注意在方法的头部使用 throws MyCustomException 声明本方法可能会抛出异常。

如果自定义异常是 RuntimeException 则无须在方法头部声明抛出。

定义好自己的异常并在某个方法中抛出后,那么在调用可能抛出自定义异常的方法的地方用 try-catch 块捕获并处理该异常,如源文件 5-15 所示。

源文件 5-15　UtilsDemo.java

```
1 public class UtilsDemo {
2     public static void main(String[] args) {
3         try {
4             System.out.println(Utils.sqrt(4));
5             System.out.println(Utils.sqrt(-4));
6         } catch (MyCustomException e) {
7             System.out.println("Caught a custom exception: "
                   + e.getMessage());
8             //或者进一步处理错误,如记录日志、显示错误消息等
9         }
10    }
11 }
```

在源文件 5-15 第 4 行使用正数 4 调用了计算平方根的自定义方法 sqrt,程序正常执行;而在第 5 行使用负数 −4 调用 Utils 类中的静态 sqrt,该方法抛出 MyCustomException 异常。该异常被捕获后,在控制台输出提示消息。注意,提示消息 error: x < 0 是在抛出异常的位置设置的(源文件 5-14 第 4 行)。程序的输出如下:

```
2.0
Caught a custom exception: error: x < 0.
```

根据图 5-4 所展示的异常处理流程,catch 块执行完毕后会去执行 finally 块和 try-catch 语句的后续语句。例如源文件 5-16 的输出结果是:

```
Begin try block
java.lang.ArithmeticException: / by zero
Finally block
End of main method
```

源文件 5-16　ProcessDemo.java

```
1 public class ProcessDemo {
2     static final int ZERO = 0;
3     public static void main(String[] args) {
4         try {
5             System.out.println("Begin try block");
6             int y = 2 / ZERO;
7             System.out.println("End try block");
8         } catch (RuntimeException e) {
9             System.out.println(e);
10        } finally {
11            System.out.println("Finally block");
12        }
13        System.out.println("End of main method");
14    }
15 }
```

所以在 catch 块中不能仅仅输出一些信息,还得要有流程控制语句以避免继续执行 try-catch 语句的后续语句。一个很常见的做法是捕获标准异常并包装为自定义异常。需要注意的是，包装异常时，一定要把原始的异常设置为 cause（Exception 有构造方法可以传入 cause）。否则，丢失了原始的异常信息会让错误的分析变得困难。

源文件 5-17　WrapDemo.java

```
 1  public class WrapDemo {
 2      static final int ZERO = 0;
 3      public static void main(String[] args) {
 4          try {
 5              wrapException();
 6          } catch (MyCustomException e) {
 7              e.printStackTrace();
 8              System.out.println(e.getMessage());
 9          }
10      }
11
12      public static void wrapException() throws MyCustomException {
13          try {
14              System.out.println("Begin try block");
15              int y = 2 / ZERO;
16              System.out.println("End try block");
17          } catch (ArithmeticException e) {
18              throw new MyCustomException("error: / by zero.", e);
19          } finally {
20              System.out.println("Finally block");
21          }
22          System.out.println("End of wrapException method");
23      }
24  }
```

在源文件 5-17 中的第 18 行把原始异常"被 0 除"包装成为自定义异常 MyCustomException，见源文件 5-13，并把原始异常作为 MyCustomException 异常的原因。这样当执行到第 15 行抛出异常时，控制会转移到第 6 行而不会继续执行第 22 行。程序输出：

```
Begin try block
Finally block
MyCustomException: error: / by zero.
    at WrapDemo.wrapException(WrapDemo.java:18)
    at WrapDemo.main(WrapDemo.java:5)
Caused by: java.lang.ArithmeticException: / by zero
    at WrapDemo.wrapException(WrapDemo.java:15)
    ... 1 more
error: / by zero.
```

其中前两行消息在 wrapException() 方法中输出；后面的消息由源文件第 7 行的 printStackTrace 方法产生；最后一行的消息由 getMessage 方法产生。

第 5 章　章节测验

第6章 实用类

Java 平台标准版(Java SE)APIs 定义了完成通用计算的类和方法,其中对字符串的操作、正规表达式、数学运算、日期和时间等是高级语言程序设计中常见的操作。完成这些操作的类,称为实用类(utility class)。

6.1 字符串

字符串就是字符序列。字符串字面量是使用引号引起来的字符序列。Java 把字符串作为 String 类型的对象。通过字面量创建字符串对象是最简单的字符串对象创建方式。例如:

```
String s;
s = "ABC";
```

编译器把字面量"ABC"编译为一个字符串对象,在运行时刻将其引用赋值给引用变量 s。此时该 String 对象及引用变量 s 的关系如图 6-1 所示。从中可以看到,字符串对象以类私有字节数组存储:private final byte[],并且一旦初始化就不允许改变,这称为字符串的不变性(immutable)。

图 6-1 字符串"ABC"

汉字"一""二"和"三"的 Unicode 代码点分别是\u4e00、\u4e8c 和\u4e09，下面的语句

```
String s, t;
s = "ABC";
t = "一二三";
```

中最后一条语句把字符串"一二三"存储在 6 字节数组中，如图 6-2 所示。字节 0 和 1 分别存放\u4e00 的低位(00)和高位(4e)；字节 2 和 3 分别存放\u4e8c 的低位(8c)和高位(4e)；字节 4 和 5 分别存放\u4e09 的低位(09)和高位(4e)。

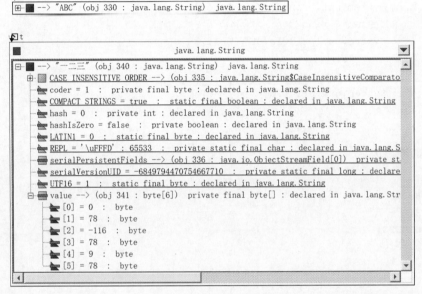

图 6-2 字符串"一二三"

除了使用字符串字面量创建字符串对象外，还可以通过无参构造方法创建字符串对象、通过字节数组创建字符串对象、通过字符数组创建字符串对象，以及通过指定字符串对象创建字符串对象。例如在源文件 6-1 中，第 5 行创建了一个长度为 0 的字符串对象。第 9 行使用字节数值创建了字符串对象；第 11 行使用字节数组中从位置 3 开始的 4 个字节创建字符串对象；而第 14 行则以字节数组和指定字符集创建字符串对象。第 18 行通过字符数组创建字符串对象，第 20 行则以从位置 3 开始的 4 个字符来创建字符串对象。第 23 行以指定字符串 abc 创建字符串对象，当然结果仍然是 abc。

源文件 6-1 StringConstructorsDemo.java

```
1 import java.nio.charset.StandardCharsets;
2 public class StringConstructorsDemo {
3     public static void main(String[] args) {
4
5         String s = new String();
6         System.out.println("s = " + s);
7
8         byte[] bytes = {65, 66, 67, 68, 69, 70, 71};
9         s = new String(bytes);
```

```
10      System.out.println("s = " + s);
11      String t = new String(bytes, 3, 4);
12      System.out.println("t = " + t);
13
14      t = new String(bytes, StandardCharsets.UTF_8);
15      System.out.println("t = " + t);
16
17      char[] chars = {'a', 'b', 'c', 'd', 'e', 'f', 'g'};
18      s = new String(chars);
19      System.out.println("s = " + s);
20      t = new String(chars, 3, 4);
21      System.out.println("t = " + t);
22
23      s = new String("abc");
24      System.out.println("s = " + s);
25   }
26 }
```

程序的输出为：

```
s =
s = ABCDEFG
t = DEFG
t = ABCDEFG
s = abcdefg
t = defg
s = abc
```

在源文件中多次出现的某个字符串字面量视为同一个字符串。例如在源文件 6-2 中的第 4 行出现了字符串"ABC"，在第 6 行也出现了字符串"ABC"，那么这两个字符串是同一个字符串，第 7 行同一性测试的输出是 true。

源文件 6-2　StringDemo.java

```
1 public class StringDemo {
2    public static void main(String[] args) {
3       String s, t;
4       s = "ABC";
5       t = "一二三";
6       String u = "ABC";
7       System.out.println(s == u);
8    }
9 }
```

图 6-3 展示了第 4 行的"ABC"是字符串对象 330；第 6 行的"ABC"也是字符串对象 330，二者是同一个对象。

java.lang.String 类是接口 java.lang.CharSequence 的实现类，也是接口 Comparable 的实现类。CharSequence 接口中定义的公共实例方法如表 6-1 所示。

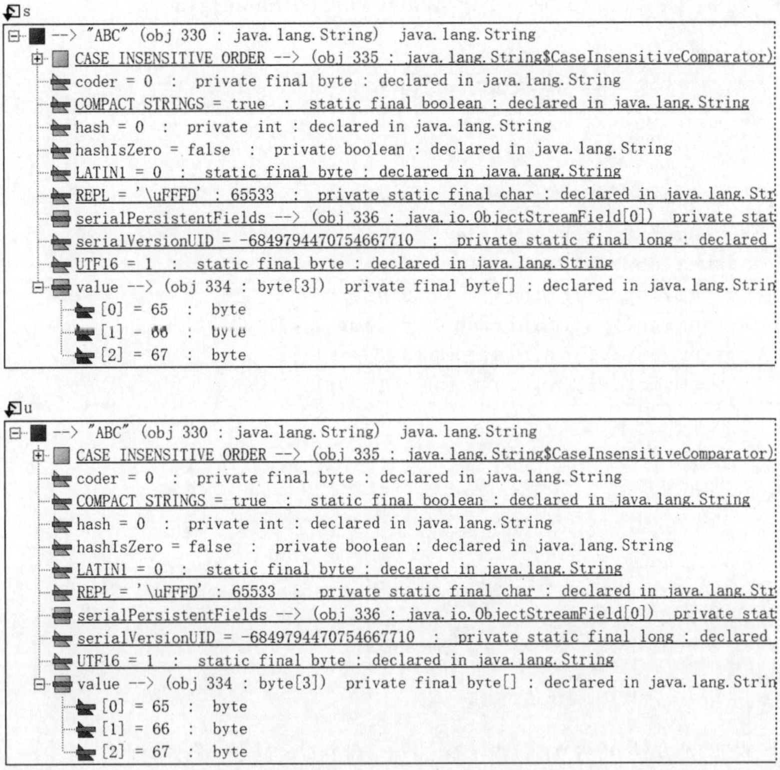

```
s
├─■ --> "ABC" (obj 330 : java.lang.String)  java.lang.String
│  ├─ CASE_INSENSITIVE_ORDER --> (obj 335 : java.lang.String$CaseInsensitiveComparator)
│  ├─ coder = 0 : private final byte : declared in java.lang.String
│  ├─ COMPACT_STRINGS = true : static final boolean : declared in java.lang.String
│  ├─ hash = 0 : private int : declared in java.lang.String
│  ├─ hashIsZero = false : private boolean : declared in java.lang.String
│  ├─ LATIN1 = 0 : static final byte : declared in java.lang.String
│  ├─ REPL = '\uFFFD' : 65533 : private static final char : declared in java.lang.Str
│  ├─ serialPersistentFields --> (obj 336 : java.io.ObjectStreamField[0])  private stat
│  ├─ serialVersionUID = -6849794470754667710 : private static final long : declared
│  ├─ UTF16 = 1 : static final byte : declared in java.lang.String
│  └─ value --> (obj 334 : byte[3])  private final byte[] : declared in java.lang.Strin
│     ├─ [0] = 65    byte
│     ├─ [1] = 66    byte
│     └─ [2] = 67    byte

u
├─■ --> "ABC" (obj 330 : java.lang.String)  java.lang.String
│  ├─ CASE_INSENSITIVE_ORDER --> (obj 335 : java.lang.String$CaseInsensitiveComparator)
│  ├─ coder = 0 : private final byte : declared in java.lang.String
│  ├─ COMPACT_STRINGS = true : static final boolean : declared in java.lang.String
│  ├─ hash = 0 : private int : declared in java.lang.String
│  ├─ hashIsZero = false : private boolean : declared in java.lang.String
│  ├─ LATIN1 = 0 : static final byte : declared in java.lang.String
│  ├─ REPL = '\uFFFD' : 65533 : private static final char : declared in java.lang.Str
│  ├─ serialPersistentFields --> (obj 336 : java.io.ObjectStreamField[0])  private stat
│  ├─ serialVersionUID = -6849794470754667710 : private static final long : declared
│  ├─ UTF16 = 1 : static final byte : declared in java.lang.String
│  └─ value --> (obj 334 : byte[3])  private final byte[] : declared in java.lang.Strin
│     ├─ [0] = 65    byte
│     ├─ [1] = 66    byte
│     └─ [2] = 67    byte
```

图 6-3 "ABC" 的两次出现

表 6-1 CharSequence 接口

返回值类型	方 法	功 能
char	charAt(int index)	返回指定位置上的字符值
int	length()	返回字符串中的字符个数
CharSequence	subSequence(int start, int end)	返回[star, end)位置上的子串
String	toString()	返回对象的字符串表示

例如源文件 6-3 中，第 7 行默认调用了 toString 方法，输出是 ABC；第 8 行查询 s 所引用字符串的长度，输出是 3；第 9 行查询 s 所引用字符串中位置 0 上的字符，输出 A；第 10 行查询 s 所引用字符串中从位置 1 开始(含 1)到位置 3(不含 3)的子串，输出为 BC。

接口 Comparable 中只定义了一个方法：int compareTo(T o)，其中 T 是类型形式参数。字符串类 String 中定义的实例方法 public int compareTo(String anotherString) 以 String 为实际参数实现了接口 int compareTo(T o)。该方法把当前字符串与另外一个字符串 anotherString 从左到右按字母表顺序逐个比较位置 i 上的字符，如果位置 i 上的字符小，则返回一个负数；如果相等，返回 0；如果大则返回一个正数。例如源文件 6-3 中的第 11 行输出字符串 s 与另外一个字符串 t 进行大小比较的结果。因为 s 中的 A 与 t 中的 A 相等，接着比较 s 中的 B 与 t 中的 B，也相等；再比较 s 中的 C 和 t 中的 D，C 小于 D，所以返回一个负数，这个负数表示 s 引用的字符串小于 t 引用的字符串。同理，第 12 行比较 ABC 和 AB，结果是一个整数 1，表示 ABC 大于 AB。

源文件 6-3　StringCharSeqDemo.java

```
1 public class StringCharSeqDemo {
2     public static void main(String[] args) {
3         String s, t, u;
4         s = "ABC";
5         t = "ABD";
6         u = "AB";
7         System.out.println(s);
8         System.out.println(s.length());
9         System.out.println(s.charAt(0));
10        System.out.println(s.subSequence(1, 3));
11        System.out.println(s.compareTo(t));
12        System.out.println(s.compareTo(u));
13    }
14 }
```

完整的程序输出为：

```
ABC
3
A
BC
-1
1
```

字符串上的实例方法可分为 4 类：查询类、判断类、转换类、分割类。

除了查询字符串的长度、指定位置的字符、指定位置的子串，还有查询子串的位置等。

在源文件 6-4 的第 4 行使用 indexOf 方法查询字符 D 在字符串中第一次出现的位置；使用 lastIndexOf 方法查询字符 D 在字符串中最后一次出现的位置；使用重载方法 indexOf 查询字符串 CD 在字符串 ABCDABCD 中第一次出现的位置；使用重载方法 lastIndexOf 查询字符串 CD 在字符串 ABCDABCD 中最后一次出现的位置。

源文件 6-4　StringQueryDemo.java

```
1 public class StringQueryDemo {
2     public static void main(String[] args) {
3         String s = "ABCDABCD";
4         System.out.println(s.indexOf('D'));
5         System.out.println(s.lastIndexOf('D'));
6         System.out.println(s.indexOf("CD"));
7         System.out.println(s.lastIndexOf("CD"));
8     }
9 }
```

程序的输出为：

```
3
7
2
6
```

判断类的方法有：判断开头或结尾的字符串、判断字符串包含、判断字符串中有无字符、判断字符串相等等。例如在源文件 6-5 中，第 5 行使用 startsWith 方法判断字符串对象

ABCD 是否以 AB 打头；第 6 行使用 endsWith 方法判断字符串 ABCD 是否以 CD 结尾；第 7 行使用 contains 方法判断字符串 ABCD 中是否包含 BC；第 8 行判断字符串 ABCD 中有无字符；第 9 行使用覆盖方法 equals 判断 ABCD 和 abcd 两个字符串是否相等；第 10 行使用 equalsIgnoreCase 方法判断 ABCD 和 abcd 在忽略大小写的情况下是否相等。

源文件 6-5   StringDecisionDemo.java

```
1 public class StringDecisionDemo {
2     public static void main(String[] args) {
3         String s = "ABCD";
4         String t = "abcd";
5         System.out.println(s.startsWith("AB"));
6         System.out.println(s.endsWith("CD"));
7         System.out.println(s.contains("BC"));
8         System.out.println(s.isEmpty());
9         System.out.println(s.equals(t));
10        System.out.println(s.equalsIgnoreCase(t));
11    }
12 }
```

程序运行结果如下：

```
true
true
true
false
false
true
```

转换类的方法有：字符替换、字符串替换、去除首尾空格、大小写转换等。例如在源文件 6-6 中，第 4 行使用 replace 方法把字符串 ABCDef 中的 A 替换为 *；第 5 行使用重载的 replace 方法把字符串 ABCDef 中的 AB 替换为 * *；第 7 行使用 trim 方法去除字符串两端空格；第 8 行使用 replace 方法去除字符串中的所有空格；第 9 行把字符串 ABCDef 中所有字母转换为小写；第 10 行则把字符串 ABCDef 中所有字母转化为大写。

源文件 6-6   StringReplaceDemo.java

```
1 public class StringReplaceDemo {
2     public static void main(String[] args) {
3         String s = "ABCDef";
4         System.out.println(s.replace('A', '*'));
5         System.out.println(s.replace("AB", "* *"));
6         String t = "   A B C D   ";
7         System.out.println(t.trim());
8         System.out.println(t.replace(" ", ""));
9         System.out.println(s.toLowerCase());
10        System.out.println(s.toUpperCase());
11    }
12 }
```

程序的输出为：

```
 * BCDef
**CDef
A B C D
ABCD
abcdef
ABCDEF
```

分割类的方法有截取子串和分割子串等。例如在源文件 6-7 中,第 6 行使用 substring 方法从字符串 9-2-302 第 5 个字符截取到末尾;第 7 行使用重载的 substring 方法从第 5 个字符截取到第 7 个字符;第 8 行将字符串 9-2-302 以"-"为分隔符切分成 9、2 和 302 三个字符串,存储在字符串数组中返回。第 9 行开始的循环语句用来输出 split 方法返回的字符串数组,如果不是数组的最后一个字符串则在后面加逗号;数组的最后一个字符串后面不加逗号。

源文件 6-7　StringSplitDemo.java

```
1 import java.util.Arrays;
2
3 public class StringSplitDemo {
4     public static void main(String[] args) {
5         String s = "9-2-302";
6         System.out.println(s.substring(4));
7         System.out.println(s.substring(4, 7));
8         String[] result = s.split("-");
9         for (int i = 0; i < result.length; i++) {
10            if (i != result.length - 1) {
11                System.out.print(result[i] + ",");
12            } else {
13                System.out.println(result[i]);
14            }
15        }
16    }
17 }
```

注意,length() 是 String 类中的方法而 length 是数组对象的属性。s.length() 用来获取字符串 s 中的字符个数,而 a.length 用来获取数组 a 中的元素个数。求子串方法 substring、转换小写方法 toLowerCase、转换大写 toUpperCase、去除首尾空格方法 trim,或者字符替换方法 replace 均返回另一个新的字符串对象,并不是更新原来的字符串对象。可以把字符串转换为字符数组,例如:

```
char[] message = "Welcome to Java".toCharArray();
```

作为 String 类应用的一个例子,源文件 6-8 演示了如何使用 String 类统计一个给定段落中的空格数和字母数。

源文件 6-8　StringParagraphDemo.java

```
1 public class StringParagraphDemo {
2     public static void main(String[] args) {
3         //被统计的段落
4         String paragraph = "The JDK includes the JRE plus command-line"
5             + "development tools such as compilers and debuggers that are"
```

```
6              + "necessary or useful for developing applets and applications.";
7          int spaces = 0;                                    //空格计数
8          int letters = 0;                                   //字母计数
9          //逐一处理段落中的每个字符
10         int paragraphLength = paragraph.length();          //段落长度
11         for (int i = 0; i < paragraphLength; i++) {
12             char ch = paragraph.charAt(i);
13             //判断是否是字母
14             if (Character.isLetter(ch)) {
15                 letters++;
16             }
17             //判断是否是空格
18             if (Character.isSpaceChar(ch)) {
19                 spaces++;
20             }
21         }
22         System.out.println("The paragraph contains" + letters + "letters and"
               + spaces + "spaces.\n");
23     }
24 }
```

程序的输出为：

```
The paragraph contains 137 letters and 23 spaces.
```

在源文件 6-8 中，首先使用字符串字面量创建被统计的字符串对象，并让 paragraph 引用该字符串对象。然后使用 for 循环逐个字符进行判断，如果是字母则对变量 letters 增 1；如果是空格，则将变量 spaces 增 1。在 for 语句之前首先使用字符串对象的 length 方法返回字符串长度。在循环体中，使用了 Character 类的静态方法 isLetter 判断是否是字母；使用 Character 类的静态方法 isSpaceChar 判断是否是空格。注意，静态方法要通过"<类名>.<方法>"的形式来调用。

Java 文本块，也称为多行字符串字面量，能够在字符串中直接使用空白符而无须转义，从而使程序更加可读。文本块使用一对连续的三个双引号（"""）作为分界符。例如源文件 6-9 输出一个由 * 组成的等腰三角形：

源文件 6-9　TextBlockExample.java

```
public class TextBlockExample {
    public static void main(String[] args) {
        String triangle = """
             *
            ***
           *****
          *******
        """;
        System.out.println(triangle);
    }
}
```

其中的文本块对象 triangle 等价于下面的字符串对象 lines：

```
String lines = "\n" +
         "   *  \n" +
         "  *** \n" +
         " *****\n" +
         "*******\n";
```

文本块对象也是 java.lang.String 对象,与通过双引号创建的字符串对象具有相同的特征。能够使用字符串字面量的地方也能够使用文本块对象。注意,文本块的开引号位于行尾;闭引号单独占一行。不能把文本块写在单独一行上:

```
//错误
String name = """<HTML></HTML>""";
```

源文件 6-10 使用文本块表示一个 HTML 页面的源代码。其输出为:

```
<HTML>
   <BODY>
       <p>天气晴朗</p>
   </BODY>
</HTML>
```

源文件 6-10　TextBlockIndent.java

```
1  public class TextBlockIndent {
2      public static void main(String[] args) {
3          String page = """
4              <HTML>
5                  <BODY>
6                      <p>天气晴朗</p>
7                  </BODY>
8              </HTML>
9              """;
10         System.out.println(page);
11     }
12 }
```

比较源文件 6-10 的输出与源程序会发现,从第 3 行到第 9 行既有用于程序设计风格的缩进(□表示空格),也有文本块中的缩进,而输出中仅包含文本块中的缩进:

```
□□□□□□□□String page = """
□□□□□□□□□□□□<HTML>
□□□□□□□□□□□□    <BODY>
□□□□□□□□□□□□        <p>天气晴朗</p>
□□□□□□□□□□□□    </BODY>
□□□□□□□□□□□□</HTML>
□□□□□□□□""";
```

识别文本块中缩进的规则是:把文本块中的所有行向左移动,直到与具有最少前导空格的行左对齐为止。所以,把<HTML>括起来的第 4 行到第 8 行左移,直到与最少前导空格的第 4 行左对齐,移过的空格就是用于文本块内缩进的空格。

6.2 正规表达式

在应用开发中,经常需要判断一个字符串是否满足一定的模式。比如邮政编码、手机号码、QQ号码都有一定的模式,应用字符串模式可以对字符串的有效性进行验证。例如判断一个给定的字符串是否仅仅由字母 A 或 B 组成。首先使用正规表达式把"仅仅由字母 A 或 B 组成"的模式串表示为:[AB]+,其中,A 和 B 是字母,[AB]含义是字母 A 或字母 B,+ 的含义是至少出现 1 次。所以,[AB]+的含义就是"字母 A 或 B 的多次出现"。[AB]+就是一个正规表达式,其特点是使用了一些约定的符号,如[]和+,对字符串模式进行描述。

Java API 中的 String 类提供了实例方法 matches(),该方法用来判断当前字符串对象是否与给定的字符串模式参数匹配。源文件 6-11 的第 5 行使用 matches()方法判断字符串 "ABBBCA"是否与模式"[AB]+"相匹配,因为字符串中含有字母 C,所以匹配失败,返回 false;第 7 行判断字符串"ABBBA"是否与模式"[AB]+"相匹配,匹配成功,返回 true。

源文件 6-11　StringRegularExp.java

```
1  public class StringRegularExp {
2      public static void main(String[] args) {
3          String regularExp = "[AB]+";
4          String inputString = "ABBBCA";
5          System.out.println(inputString.matches(regularExp));
6          inputString = "ABBBA";
7          System.out.println(inputString.matches(regularExp));
8      }
9  }
```

程序的输出为:

```
false
true
```

除了 matches 方法外,String 类还提供了 split、replaceFirst、replaceAll 等可以使用正规表达式的实例方法:split 方法根据给定正规表达式切分字符串;replaceFirst 方法替换与给定正规表达式匹配的第 1 个子串;replaceAll 方法替换与给定正规表达式匹配的所有子串。

源文件 6-12 把字符串"one:two::three:::four::::five"以"若干个冒号"为分隔符分割为若干个字符串。第 5 行以模式串":+"为 split 方法的实际参数,返回 5 个字符串:one、two、three、four、five 构成的字符串数组。for-each 循环逐个输出字符串数组中的各个子串。

源文件 6-12　SplitDemo.java

```
1  public class SplitDemo {
2      public static void main(String[] args) {
3          String regularExp = ":+";
4          String inputString = "one:two::three:::four::::five";
5          for (String s : inputString.split(regularExp))
```

```
6            System.out.println(s);
7        }
8    }
```

前文介绍了正规表达式由普通字符（字母、数字等）和一些具有特定含义的字符组成。比如"+"的含义是"出现至少1次"。这些预先规定的具有特殊含义的字符称为"元字符（meta-character）"。表示出现次数的元字符有?、*、+和{}。这些符号的具体含义如表6-2所示。表中的X表示任意的普通字符序列。

表6-2　表示出现次数的元字符

字　符	含　义	字　符	含　义
X?	X出现0次或1次	X{n}	X重复出现n次
X*	X出现0次或多次	X{n,}	X重复出现至少n次
X+	X出现1次或多次	X{n,m}	X重复出现至少n次,至多m次

假设符号串"aaaaa"由变量s引用,表6-3列举了几个正规表达式的例子,解释这些特殊符号的含义。

表6-3　出现次数表示示例

正规表达式	s.matches()匹配结果	正规表达式	s.matches()匹配结果
a?	false	a{3}	false
a*	true	a{3}	true
a+	true	a{3,5}	true

表6-3中的正规表达式仅仅是对单个字符"a"进行重复次数的描述。想表达字符串"abababab"的模式,即描述对字符串"ab"重复模式,比如,是否写成"ab+"呢？"ab+"的含义是"以字符a打头,后面至少出现1个字符b"。比如"abbb"。"(ab)+"的含义是"至少出现1次ab"。所以,重复出现的字符序列不止1个字符时,应使用圆括号()括起来。

正规表达式（abc|xyz）+的含义是"abc或者xyz至少出现1次"。那么abc、xyz、abcabc、abcxyz都满足该模式。而ab、xy、abyz则不满足该模式。

假设X,Y是正规表达式,那么XY的含义是X的后面是Y；X|Y意思是要么X,要么Y。

如何描述"数字"这样的"一类字符"呢？按照上面介绍的写法,可以写成：

(0|1|2|3|4|5|6|7|8|9)+

那么非0打头的数字字符串呢？就得写成：

(1|2|3|4|5|6|7|8|9)(0|1|2|3|4|5|6|7|8|9)*

那么,如何描述"大写字母组成的字符串"呢？是否把所有26个大写字母都列出来呢？使用特殊符号()和|表达这样的模式就显得费劲了。为了解决这个问题,引入了描述字符类的特殊符号。

特殊符号[]表示"或"运算,枚举在特殊符号[]中的字符即运算对象。例如,[abc]含义是a或b或c。相当于写成：

(a|b|c)

还有特殊符号^和-,其含义如表6-4所示。

表6-4 正规表达式中的特殊符号

符 号	举 例	含 义
[]	[abc]	a 或 b 或 c
^	[^abc]	除了 a、b、c 以外的其他符号
-	[a-zA-Z]	指定范围,从 a 到 z,或者从 A 到 Z

有了这三个特殊符号,就可以比较简洁地描述"一类"字符。例如,描述"大写字母组成的字符串",模式串可以是:

[A-Z]+

描述"任意字母"的模式串就可以是:

[a-zA-Z]

描述"数字串"的正规表达式就可以是:

[0-9]+

描述"非0打头的数字串"的正规表达式就可以是:

[1-9][0-9]*

可以发现,这些表达式比前文中的表达式简洁了很多。

正规表达式 [bcr]at 的含义是"以字母 b 或 c 或 r 打头,后随字母 a 和 t",所以字符串 "bat" "cat" 以及 "rat" 匹配该模式。正规表达式 [^bcr]at 的含义是"不以字母 b、c 和 r 打头,后随字母 a 和 t",所以字符串"hat"匹配该模式。

正规表达式 [0-4[6-9]] 的含义等同于[012346789],但是看起来更容易理解。

为了方便使用,Java 还预定义了一些常用的"字符类"的记号,如表6-5所示。

表6-5 预定义的字符类

符号	字 符 类
.	任意字符
\d	数字:[0-9]
\D	非数字:[^0-9]
\s	空白字符(whitespace character):[\x0B \t \f \n \r]
\S	非空白字符[^\s]
\w	单词字符(word character):[a-zA-Z_0-9]
\W	非单词字符(non-word character):[^\w]

有了这些定义,使得模式描述更加简洁。比如"数字串"的正规表达式就可以是:

\d+

^和$分别匹配字符串的开始和结束。例如要求密码长度为8~16位,而且只允许包含

大写字母、小写字母和数字。那么就可以使用如下正规表达式：

^[a-zA-Z\d]{8,16}$

其中，^表示匹配字符串的开始；[a-zA-Z\d]{8,16}表示密码由大小写字母和数字组成，且长度为8~16位；$表示匹配字符串的结束。

表6-6列举了几个常用的正规表达式，通常用于对用户输入的校验。

表6-6 常用的正规表达式

正规表达式	匹配
[1-9][0-9]{3,9}	QQ账号
^(\s*)\r\n	空行
[\u4e00-\u9fa5]+	一个或多个汉字
[A-Za-z]+	仅包含英文字母的字符串
[0-9]+	数字串
\d{n}	长度为n的数字串
\d{n,}	长度至少为n的数字串
\d{n,m}	长度至少为n、至多为m的数字串
+?[1-9]\d*	正整数
[0-9]+(\.[0-9]{2})?	小数点后只有2位的数字串

在split、replaceFirst、replaceAll等方法中应用正规表达式使得对字符串的处理更加灵活、简洁。例如源文件6-13试图删除字符串中的所有空白符号，那么就应用正规表达式"\s+"表示若干个空白符，应用replaceAll方法把匹配该模式的所有子串替换为空字符串。

源文件6-13 **StringRegularExpDemo.java**

```
1 public class StringRegularExpDemo {
2     public static void main(String[] args) {
3         String s = "   A B C D   ";
4         System.out.println(s.replaceAll("\\s+", ""));
5     }
6 }
```

程序的输出为：

ABCD

注意：空格（space）、制表（tabulation）、回车（carriage return）、换行（line feed）、换页（form feed）都是空白符（white space）。还需要注意，源文件6-13的第4行replaceAll方法的第一个实际参数是正规表达式字符串"\\s+"，而正规表达式是"\s+"。这是因为在字符串字面量中，反斜线"\"表示"转义"，比如"\t"表示制表符而不是\和t两个字母。所以在\前面再写一个\，把后面的\转义为Java字符串字面量中的普通字符。但在正规表达式\s+中，\仍然解释为"转义"。

如果需要从字符串中查找与指定模式匹配的多个子串，则需要使用java.util.regex.Matcher类。Matcher类是Java正规表达式API中的一个重要组成部分，与Pattern类配套使用，负责执行实际的匹配操作。一般首先把某个正规表达式r编译为Pattern对象，然后以被查找的字符串为实际参数,从该 Pattern 对象创建 Matcher 对象。这个 Matcher 对象的实例方法 find 查找输入字符串中是否存在匹配正规表达式 r 的子串。如果找到匹配项,返回 true;否则返回 false。实例方法 group 获取整个正规表达式匹配的下一个子串。如果没有匹配项,则抛出异常。通常使用一个循环语句遍历所有的匹配项。源文件 6-14 展示了如何使用 Matcher 类从给定字符串中查找所有电话号码。

源文件 6-14　MatcherDemo.java

```
1 import java.util.regex.Pattern;
2 import java.util.regex.Matcher;
3
4 public class MatcherDemo {
5     public static void main(String[] args) {
6         String s = "张三 010-66001234;李四 0311-80781234;王五 88001234";
7         Pattern pattern = Pattern.compile("(0\\d{2,3}-\\d{8})|\\d{8}");
8         Matcher matcher = pattern.matcher(s);
9         while (matcher.find()) {
10             System.out.println(matcher.group());
11         } //end while
12     } //end main
13 }
```

程序的输出为:

```
66001234
0311-80781234
88001234
```

源文件 6-14 中的第 6 行使用字符串字面量创建了一个含有三个电话号码的字符串对象。有的电话号码含有区号,并使用了分隔符-。有的电话号码没有区号。所有不含区号的电话号码都是 8 位数字,所以可以使用正规表达式\d{8}表示这种模式;每个区号都是 0 打头,后面至少 2 个数字,至多 3 个数字,所以可使用正规表达式 0\d{2,3}表示这种模式。因此正规表达式 0\d{2,3}-\d{8}就表示了含区号的电话号码。那么含区号或不含区号的电话号码的正规表达式就是 0\d{2,3}-\d{8}|\d{8}。第 7 行就使用这个正规表达式创建了 Pattern 对象。或者使用正规表达式 0\d{2,3}-? \d{8}也可以。

第 8 行从 Pattern 对象上生成 Matcher 对象来实现匹配功能,并指定输入字符串。第 9 行开始的循环用来遍历输出所有的匹配项。

在正规表达式圆括号()内的子表达式称为 group,group 按顺序索引,可以嵌套。默认整个正规表达式是一个 group。所以,无参数的方法调用 group()返回与整个表达式匹配的下一个字符串。源文件 6-15 从<li>2023-04-26</li>中查找与正规表达式 (\d{4})-(\d\d)-(\d\d)匹配的所有子串。在第 11 行使用 group 方法返回与整个正规表达式匹配的子串;第 12 行到第 14 行分别使用组号 1、2、3 返回正规表达式中指定组的匹配项。第 1 个组是\d{4};第 2 个组是中间使用括号括起来的\d\d;第 3 个组是末尾使用括号括起来的\d\d。

源文件 6-15　GroupDemo.java

```
1 import java.util.regex.Pattern;
2 import java.util.regex.Matcher;
```

```
3
4 public class GroupDemo {
5     public static void main(String[] args) {
6         String dataSource = "<li><span>2023-04-26</span></li>";
7         String regex = "(\\d{4})-(\\d\\d)-(\\d\\d)";
8         Pattern p = Pattern.compile(regex);
9         Matcher m = p.matcher(dataSource);
10        while (m.find()) {
11            System.out.println(m.group());
12            System.out.print(m.group(1) + ", ");
13            System.out.print(m.group(2) + ", ");
14            System.out.println(m.group(3));
15        }
16    }
17 }
```

程序的输出为：

```
2023-04-26
2023, 04, 26
```

6.3 编辑字符串

一般使用 String 类完成字符串的查询、判断、转换、分割等操作。但是转换后的结果是一个新的字符串对象，并不能改变原来的字符串对象。这是因为一旦创建了字符串对象，就不能改动其内容。在如下代码中：

```
String s = "Hello";
s = s + " everyone!";
```

s 原先指向一个 String 对象，内容是 "Hello"，然后对 s 进行了字符串连接操作，这时，s 不指向原来那个对象了，而指向了另一个 String 对象，内容为"Hello everyone!"，原来那个对象还存在于内存之中，只是 s 这个引用变量不再指向它了。

如果需要对一个字符序列频繁地增加字符、删除字符或者替换字符，那么使用 String 来处理字符串会引起很大的内存开销。这种情况适合使用 StringBuffer 类。例如如果要修改已有的字符串 Hello 为 Hello everyone!，那么可以通过在 StringBuffer 对象上应用方法 append 实现追加：

```
StringBuffer s = new StringBuffer("Hello");
s.append(" everyone!");
```

源文件 6-16 演示了对 StringBuffer 对象追加、插入、替换等操作。所有操作均在同一个存储空间上进行。

源文件 6-16　StringBufferDemo.java

```
1 public class StringBufferDemo {
2     public static void main(String[] args) {
3         StringBuffer sb = new StringBuffer("Hello");
```

```
4       sb.append(" everyone!");
5       System.out.println(sb);
6
7       sb.insert(5, ",");
8       System.out.println(sb);
9
10      sb.setCharAt(5, '!');
11      System.out.println(sb);
12      sb.replace(0, 5, "Hi");
13      System.out.println(sb);
14      System.out.println(sb.reverse());
15
16      sb.reverse();
17      sb.deleteCharAt(2);
18      System.out.println(sb);
19      sb.delete(0, sb.length());
20      System.out.println(sb);
21   }
22 }
```

程序的输出为:

```
Hello everyone!
Hello, everyone!
Hello! everyone!
Hi! everyone!
! enoyreve ! iH
Hi everyone!
```

第 3 行首先以字符串 Hello 为构造方法的参数创建 StringBuffer 字符串对象,接着第 4 行使用 append 方法在字符串末尾添加字符串" everyone!"。第 7 行在位置 5 后面插入字符串。

第 10 行把位置 5 上的字符","改为"!",第 12 行把[0,5)位置上的字符 Hello 替换为 Hi。第 14 行反转字符串。第 16 行再反转。

第 17 行删除位置 2 上的字符"!"。第 19 行删除[0,12)上的字符。当前字符串的长度为 12,即 sb.length()返回 12。length 方法返回 StringBuffer 对象中实际存储的字符个数。删除所有的字符,即清空字符串缓冲区。

StringBuffer 类的 toString 方法用来得到一个字符序列相同的字符串对象。capacity 方法返回 StringBuffer 对象能够存储的字符个数,即其容量。默认的初始容量是 16。当追加字符串时,容量就会动态增长。

String 和 StringBuffer 都可以存储和操作字符串,都实现了 CharSequence 接口和 Comparable 接口。不同之处在于,String 类用来创建内容不可改变的字符串;而 StringBuffer 类用来创建内容可以被编辑的字符串。String 对象可以应用方法 equals;而 StringBuffer 不能。

StringBuffer 是线程安全的,适用于多线程环境。在单线程应用中建议使用与之兼容的 StringBuilder 类。

6.4 字符对象

char 是一种基本数据类型，源文件 6-17 中的字面量'a'、'\u03B1'、'A'等都是值而不是对象。

源文件 6-17　CharacterDemo.java

```
1 public class CharacterDemo {
2     public static void main(String[] args) {
3         char ch = 'a';
4         //大写的希腊字母α的Unicode代码点
5         char alpha = '\u03B1';
6         char[] charArray = {'A', 'B', 'C', 'D'};
7         Character obj = 'a';
8         System.out.println((char)(ch + 1));
9         System.out.println((char)(obj + 1));
10        System.out.println(Character.isLetter(ch));
11    }
12 }
```

很多情况下需要把一个字符作为对象来使用，这个时候就需要一种方便的方式把字符类型的值包装成对象。相应于 char 类型，Java 提供了包装类 Character。其他基本数据类型也有相应的包装类，如图 6-4 所示。

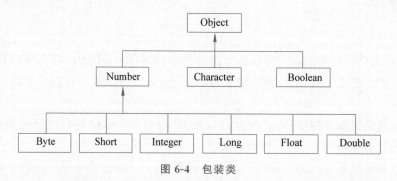

图 6-4　包装类

可以看到，这 8 种包装类中，Byte、Short、Integer、Long、Float 和 Double 都是 Number 类。使用构造方法从字符字面量构造字符对象，例如：

```
Character obj = new Character('a');
```

也可以如源文件 6-17 第 7 行那样简单地写成：

```
Character obj = 'a';
```

obj 是一个引用类型的变量，而'a'是一个 char 类型的值，二者类型并不兼容。但是编译器把这种写法看作是一种速记形式，由编译器负责从字符类型的值'a'来创建 Character 类型的对象，并把这个动作称为"自动装箱（Autoboxing）"。反之，从 Character 类型的对象转换为基本数据类型的值称为"自动拆箱（Unboxing）"，如第 9 行所示。在第 9 行，表达式 obj+1 把一个引用类型的对象与整数 1 相加，显然类型不兼容。但是编译器负责把 Character 类型的对象转换为 char 类型的值，这样就类型兼容了。

第 10 行使用了类方法 isLetter 判断字符 ch 是否为字母。其他的类方法还有：boolean isDigit(char ch)，boolean isWhitespace(char ch)，boolean isUpperCase(char ch)，boolean isLowerCase(char ch)，char toUpperCase(char ch)，char toLowerCase(char ch)等。由于这些方法是静态的，所以在调用该方法时要在前面添加类名，并使用成员访问运算符"."。

注意：Character 对象是不可变的，因此一旦创建 Character 对象，就不能再修改。

6.5 数值对象

Byte、Short、Integer、Long、Float 和 Double 都是 Number 类。Number 类的实例称为数值对象。以包装类 Integer 为例，Integer 包装类中提供对整数操作的方法，常见方法如图 6-5 所示。Number 类中有方法 doubleValue、floatValue、intValue、longValue、shortValue 和 byteValue。这些方法把对象转换成基本数据类型的数值。由于 Integer 类的对象也是 Number，所以这些方法在 Integer 对象上也可以使用。

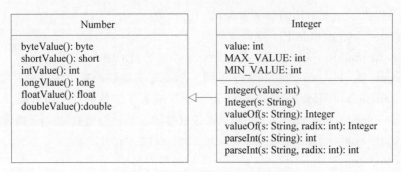

图 6-5　Integer 类中的常用方法

Integer 类中的常量 MAX_VALUE 和 MIN_VALUE 存储 int 数据类型的最大值和最小值。

Integer 类中有两个构造方法，分别以 int 类型的值或者字符串作为参数创建 Integer 对象。例如：

```
Integer x = new Integer(23);
```

假设 x 为 int 类型，对于基本数据类型的整数字面量 23，通过赋值表达式 x=23 就直接将其存储到了变量 x 中。而如果使用 Integer 对象，则首先以字面量 23 作为参数创建 Integer 对象，通过引用变量来访问这个对象。也可以自动装箱和自动拆箱：

```
Integer x = 23;
int i = x;
```

Integer 类的 valueOf 方法从数字串创建数值对象，例如：

```
Integer y = Integer.valueOf("23");
```

还有一个重载方法 valueOf(String s, int radix)可以按指定数值的进制创建数值对象。Integer 包装类的 parseInt 方法用以把字符串转换为数值。例如：

```
intn = Integer.parseInt("1234");
```

由于 parseInt 方法是静态的,所以在调用该方法时要在前面添加类名,并使用成员访问运算符"."。

表达式 Integer.parseInt("1234")的意思是:以字符串"1234"作为参数,调用 Integer 类中的方法 parseInt,将字符串转换为整数并返回该整数。

Integer 的类方法 toString(int i, int radix)把整数 i 按照指定的进制 radix 转换为数字串。

6.6 数学运算 API

6.6.1 Math 类

java.lang.Math 类包含用于执行基本数学运算的方法,如指数、对数、平方根和三角函数。类似这样的工具类,其所有方法均为静态方法,不用创建对象,调用起来非常简单。除此之外该类还提供了一些常用的数学常量,如 PI、E 等。

public static final double PI:返回圆周率。
public static double abs(double a):返回 double 类型的绝对值。
public static double ceil(double a):返回大于或等于参数的最小的整数。
public static double floor(double a):返回小于或等于参数的最大的整数。
public static long round(double a):返回最接近参数的 long 型整数(相当于四舍五入方法)。
public static double pow(double a,double b):返回 a 的 b 次幂。
public static double sqrt(double a):返回 a 的平方根。
public static double random():返回[0,1)中的随机值。
public static double max(double x, double y):返回 x,y 中的较大值。
public static double min(double x, double y):返回 x,y 中的较小值。

源文件 6-18 演示了如何使用数学运算 API。第 4 行访问了 Math 类中声明的常量 PI,即数学中的 π。

源文件 6-18　MathDemo.java

```
1 public class MathDemo {
2     public static void main(String[] args) {
3         //返回圆周率
4         System.out.println("PI = " + Math.PI);
5
6         //绝对值
7         System.out.println("abs = " + Math.abs(-25));
8
9         //大于或等于参数的最小整数,小于或等于参数的最大整数
10        System.out.println("ceil = " + Math.ceil(3.14));
11        System.out.println("floor = " + Math.floor(3.14));
12        System.out.println("ceil = " + Math.ceil(-3.14));
```

```
13          System.out.println("floor = " + Math.floor(-3.14));14
15          //四舍五入
16          System.out.println("round(9.85) = " + Math.round(9.85));
17          System.out.println("round(2.11) = " + Math.round(2.11));
18
19          //a 的 b 次幂
20          System.out.println("pow = " + Math.pow(2, 10));
21
22          //平方根
23          System.out.println("sqrt = " + Math.sqrt(81));
24
25          //[0,1)中的随机值
26          System.out.println("random = " + Math.random());
27
28          //x,y 中的较大值
29          System.out.println("max = " + Math.max(55, 66));
30
31          //x,y 中的较小值
32          System.out.println("min = " + Math.min(55, 66));
33
34      }
35  }
```

程序的输出为：

```
PI = 3.141592653589793
abs = 25
ceil = 4.0
floor = 3.0
ceil = -3.0
floor = -4.0
round(9.85) = 10
round(2.11) = 2
pow = 1024.0
sqrt = 9.0
random = 0.7973478081256746
max = 66
min = 55
```

6.6.2 Random 类

Java 实用工具类库中的类 java.util.Random 提供了各种实例方法用来产生 int、long、float、double 等类型的随机数。而 java.lang.Math 中的方法 random()只产生 double 型的随机数。

该类上生成随机数的方法有：方法调用 nextInt()返回 int 类型随机数；方法调用 nextInt(int n)返回 int 类型随机数,范围：[0,n)；方法调用 nextDouble()返回 double 类型随机数,范围：[0.0,1.0)；方法调用 nextBoolean()返回 boolean 类型随机数（true/false）。源文件 6-19 演示了如何生成随机数。

源文件 6-19 RandomDemo.java

```java
1 import java.util.Random;
2
3 public class RandomDemo {
4     public static void main(String[] args) {
5         //使用 java.lang.Math 的 random 方法生成随机数
6         System.out.println("Math.random(): " + Math.random());
7
8         Random r = new Random();
9
10        //按均匀分布产生整数
11        System.out.println("int: " + r.nextInt());
12
13        //按均匀分布产生长整数
14        System.out.println("long: " + r.nextLong());
15
16        //按均匀分布产生大于或等于0,小于1的float数[0, 1)
17        System.out.println("float: " + r.nextFloat());
18
19        //按均匀分布产生[0, 1)范围的double数
20        System.out.println("double: " + r.nextDouble());
21
22        //按正态分布产生随机数
23        System.out.println("Gaussian: " + r.nextGaussian());
24
25        //Random的nextInt(int n)方法返回一个[0, n)范围内的随机数
26        System.out.println(r.nextInt(10));
27
28        //[MIN, MAX)范围内的随机整数
29        //因为nextInt(int n)方法返回值范围是[0, n),
30        //所以需要把区间[MIN, MAX)转换成 MIN + [0, MAX - MIN + 1)。
31        int MAX = 100, MIN = 0;
32        System.out.print(MIN + r.nextInt(MAX - MIN + 1));
33    }
34 }
```

程序的输出为：

Math.random(): 0.493768067825461
int: 859033594
long: 2925129956669922965
float: 0.59081244
double: 0.020080652695474543
Gaussian: -1.224967924484732
4
57

再次运行,程序的输出结果为：

Math.random(): 0.23377317858085522
int: 1180868252
long: -6662724836192862735
float: 0.3305931
double: 0.27358439582546157

```
Gaussian: 0.6919137763284277
9
43
```

默认 Random 对象使用当前系统时钟值(以毫秒为单位)作为种子(seed)。由于系统时钟总是在变化,所以每次方法调用所生成的随机数序列是不同的。如果把种子固定到一个数,那么每次运行程序所得到的随机数序列是相同的:

```
Random aRandom = new Random();
//设置种子
aRandom.setSeed(7);          //每次运行都是同一个随机数
System.out.println(aRandom.nextInt(10));
```

运行,输出为:

```
9
```

再次运行,输出仍然是:

```
9
```

6.6.3　BigInteger 类

即使使用 long 数据类型,能处理的最大整数也只是 $2^{63}-1$。如果需要处理更大的整数,则需要使用 java.math.BigInteger。BigInteger 对象能够进行任意位数的整数的算术运算。一般首先使用字符串表示的整数来创建大整数对象,然后使用 BigInteger 类上的加方法 add、减方法 subtract、乘方法 multiply、除方法 divide、取余方法 remainder、取反方法 negate 等进行算术运算。源文件 6-20 演示了溢出现象以及 BigInteger 类的用法。

源文件 6-20　BigIntegerDemo.java

```
1  import java.math.BigInteger;
2
3  public class BigIntegerDemo {
4      public static void main(String[] args) {
5          System.out.println(Long.MAX_VALUE);
6          System.out.println(Long.MAX_VALUE + 1);
7
8          BigInteger a = new BigInteger("9223372036854775807");
9          BigInteger b = new BigInteger("1");
10         //计算 a + b
11         System.out.println(a.add(b));
12     }
13 }
```

程序的输出为:

```
9223372036854775807
-9223372036854775808
9223372036854775808
```

源文件 6-20 的第 5 行输出最大的 long 类型的值,第 6 行把该值再加 1,结果成为一个负数。第 8 行把最大的 long 型值作为字符串实参创建为 BigInteger 对象,然后在第 11 行使用 add 方法与另外一个值为 1 的 BigInteger 对象相加,得到了预期的结果。

更为复杂的四则运算,例如:

(6666666666666666 + 2222222222222222) ÷ 2

也可以使用 BigInteger 实现。首先把操作数以字符串类型的字面量参数创建 BigInteger 对象。分别使用 a 和 b 引用。然后让对象 a 执行 add 方法与对象 b 相加,相加的结果是一个新的 BigInteger 对象。通过 c 来引用。

```
BigInteger a = new BigInteger("6666666666666666");
BigInteger b = new BigInteger("2222222222222222");
BigInteger c;
//计算 a + b
c = a.add(b);
//计算 (a + b) /2
BigInteger average = c.divide(new BigInteger("2"));
//输出结果
System.out.println("Average is = " + average);
```

把 2 作为字符串字面量创建值为"2"的 BigInteger 对象,这样使得 c 能够执行 divide 方法除以 2。

除了实例方法 add、divide 外,其他实例方法还有:subtract(BigInteger val)返回当前大整数对象与参数指定的大整数对象的差;multiply(BigInteger val)返回当前大整数对象与参数指定的大整数对象的积;remainder(BigInteger val)返回当前大整数对象与参数指定的大整数对象的余数;compareTo(BigInteger val)返回当前大整数对象与参数指定的大整数的比较结果,返回值是 1、-1 或 0,分别表示当前大整数对象大于、小于或等于参数指定的大整数;pow(int a)返回当前大整数对象的 a 次幂;toString()返回当前大整数对象十进制的字符串表示;toString(int radix)返回当前大整数对象 radix 进制的字符串表示。

6.7 日期和时间

年、月、日形成了日期,时、分、秒和纳秒构成了时间。在 Java 8 及更高版本中,处理日期和时间的 API 有了显著的改进,主要集中在 java.time 包中。如 java.time.LocalDate 建模仅包含年、月、日的日期,不包含时间信息。java.time.LocalTime 建模包含时、分、秒、纳秒的时间,不包含日期。java.time.LocalDateTime 是结合了日期和时间的类。java.time.ZonedDateTime 则包含日期、时间以及时区信息。java.time.Period 建模两个日期之间的距离。java.time.Duration 表示两个时间点之间的距离。java.time.DateTimeFormatter 则用于日期和时间的格式化和解析。以下介绍这些类及其示例。

6.7.1 LocalDate 类

一个日期由年、月和日三个分量组成。java.time.LocalDate 类表示不带时区信息的日期,不含时间。可以通过静态方法 of 按指定年月日来创建 LocalDate 对象,也可以用静态方法 parse 从一个日期字符串来创建 LocalDate 对象,或者使用 now()方法创建 LocalDate 对象。

LocalDate 还提供了一些访问日期的年、月、日分量的常用方法，判断日期的先后顺序方法，修改日期的年、月、日等信息的方法，格式化日期为指定的字符串形式的方法等。源文件 6-21 演示了 LocalDate 类的使用。

源文件 6-21　LocalDateDemo.java

```java
1  import java.time.LocalDate;
2  import java.time.format.DateTimeFormatter;
3
4  public class LocalDateDemo {
5      public static void main(String[] args) {
6          //默认格式要求的日期字符串
7          String s = "1999-02-23";
8          //把日期字符串解析成日期对象
9          LocalDate birthday = LocalDate.parse(s);
10         //增加 1 年
11         System.out.println(birthday.plusYears(1));
12         //减 10 天
13         System.out.println(birthday.minusDays(10));
14         //将 LocalDate 实例设置为 1998 年 5 月 2 日
15         LocalDate modified = birthday.withYear(1998).withMonth(5)
16             .withDayOfMonth (2);
17         System.out.println(modified);
18
19         LocalDate firstDay = LocalDate.of(2024, 2, 25);
20         //访问 LocalDate 实例的分量年
21         System.out.println(firstDay.getYear());
22         //访问 LocalDate 实例的分量月
23         System.out.println(firstDay.getMonthValue());
24         //访问 LocalDate 实例的分量天
25         System.out.println(firstDay.getDayOfMonth());
26         //按 yyyy 年 MM 月 dd 日格式输出
27         System.out.println(firstDay.format(DateTimeFormatter
28             .ofPattern("yyyy年MM月dd日")));
29
30         LocalDate now = LocalDate.now();
31         System.out.println(now);
32         //判断日期 firstDay 是否在 now 之前
33         System.out.println(firstDay.isBefore(now));
34         //判断日期 firstDay 是否在 now 之后
35         System.out.println(firstDay.isAfter(now));
36         //判断日期 firstDay 和 now 是否相等
37         System.out.println(now.equals(firstDay));
38         //判断日期 firstDay 是否是闰年
39         System.out.println(firstDay.isLeapYear());
40     }
41 }
```

程序的输出为：

```
2000-02-23
1999-02-13
1998-05-02
```

```
2024
2
25
2024年02月25日
2024-01-25
false
true
false
true
```

6.7.2 LocalTime 类与 LocalDateTime 类

时间由时、分、秒等分量组成。java.time.LocalTime 类用来创建纳秒精度的时间对象。在国际单位制中,纳秒是时间的基本单位秒的下一级单位,1 纳秒等于十亿分之一秒。1 纳秒=1ns=0.000000001 秒=10^{-9} 秒。与 java.time.LocalDate 类似,也可以通过静态方法 of、静态方法 parse 以及静态方法 now 创建时间对象。LocalTime 类中提供了获取时间对象的方法,也提供了时间格式化、增减时分秒等常用方法,这些方法与日期类相对应。源文件 6-22展示了 LocalTime 类的用法。

源文件 6-22　LocalTimeDemo.java

```
1  import java.time.LocalTime;
2  import java.time.format.DateTimeFormatter;
3
4  public class LocalTimeDemo {
5      public static void main(String[] args) {
6          //获取当前时间,包含纳秒数
7          LocalTime now = LocalTime.now();
8          System.out.println(now);
9          //将时间字符串解析为时间对象
10         System.out.println(LocalTime.parse("06:15:30"));
11
12         LocalTime departureTime = LocalTime.of(19, 23, 45);
13         //访问分量:时
14         System.out.println(departureTime.getHour());
15         //按 HH 点 mm 分 ss 秒格式输出
16         System.out.println(departureTime.format(DateTimeFormatter
17             .ofPattern("HH点mm分ss秒")));
18         //判断时间前后
19         System.out.println(now.isBefore(departureTime));
20
21         //从 LocalTime 获取当前时间,不包含纳秒数
22         System.out.println(now.withNano(0));
23     }
24 }
```

运行该程序,执行结果是:

```
18:04:29.498694200
06:15:30
19
19点23分45秒
```

```
true
18:04:29
```

源文件 6-22 的第 8 行输出了含纳秒的时间对象：18:04:29.498694200，其中 498694200 是纳秒数，表示晚上 6 点 4 分 29 秒 498694200 纳秒。注意，使用 parse 方法解析字符串时，该字符串要形如"06:15:30"而不能为"6:15:30"。

第 10 行和第 12 行分别展示了如何使用静态方法 parse 和类方法 of 创建时间对象。后面与对日期对象的操作类似，展示了访问时间分量、格式化输出、判断前后的示例。

第 22 行创建了不含纳秒数的时间对象。

LocalDateTime 类是 LocalDate 类与 LocalTime 类的综合，既包含日期也包含时间。LocalDateTime 类中的方法包含了 LocalDate 类与 LocalTime 类的方法。需要注意的是，LocalDateTime 默认的格式形如"202002-29T212326.774"，这与人们经常使用的格式不太符合，所以它经常与 DateTimeFomatter 一起使用指定格式。除了 LocalDate 与 LocalTime 类中的方法外，还额外提供了转换的方法。源文件 6-23 展示了 LocalDateTime 类的用法。

源文件 6-23　LocalDateTimeDemo.java

```
1 import java.time.LocalDateTime;
2 import java.time.format.DateTimeFormatter;
3
4 public class LocalDateTimeDemo {
5     public static void main(String[] args) {
6         //系统当前日期时间
7         LocalDateTime now = LocalDateTime.now();
8         System.out.println(now);
9         //将目标 LocalDateTime 转换为相应的 LocalDate 实例
10        System.out.println(now.toLocalDate());
11        //将目标 LocalDateTime 转换为相应的 LocalTime 实例
12        System.out.println(now.toLocalTime());
13        //按照指定格式输出
14        DateTimeFormatter ofPattern = DateTimeFormatter
15            .ofPattern("yyyy年 MM月 dd日 HH时 mm分 ss秒");
16        System.out.println(now.format(ofPattern));
17    }
18 }
```

程序的输出为：

```
2024-01-25T18:26:07.016492700
2024-01-25
18:26:07.016492700
2024 年 01 月 25 日 18 时 26 分 07 秒
```

两个日期之间的距离称为期间(Period)，java.time.Period 以年、月和天为单位建模日期间的距离。两个时间之间的距离称为间隔(Duration)，java.time.Duration 以秒和纳秒为单位建模时间之间的距离。源文件 6-24 演示了 Period 类和 Duration 类的用法。

源文件 6-24　PeriodDemo.java

```
1 import java.time.LocalDate;
2 import java.time.Period;
3 import java.time.Duration;
```

```
 4 import java.time.LocalTime;
 5
 6 public class PeriodDemo {
 7     public static void main(String[] args) {
 8         LocalDate birthday = LocalDate.of(1999, 12, 23);
 9         LocalDate now = LocalDate.now();
10         //计算两个日期间的距离
11         Period between = Period.between(birthday, now);
12         System.out.println(between);
13         System.out.println(between.getYears() + "岁");
14         System.out.println(between.getMonths() + "月");
15         System.out.println(between.getDays() + "天");
16
17         LocalTime start = LocalTime.now();
18         LocalTime end = LocalTime.of(21, 34 ,56);
19         //计算两个时间之间的距离
20         Duration duration = Duration.between(start, end);
21         System.out.println(duration);
22
23         System.out.println(duration.toHours() + "小时");
24         System.out.println(duration.toMinutes() + "分");
25         System.out.println(duration.toSeconds() + "秒");
26         System.out.println(duration.toMillis() + "毫秒");
27         System.out.println(duration.toNanos() + "纳秒");
28     }
29 }
```

程序的输出为：

```
P24Y1M2D
24 岁
1 月
2 天
PT2H40M53.0566316S
2 小时
160 分
9653 秒
9653056 毫秒
9653056631600 纳秒
```

源文件 6-24 的第 11 行计算了生日和当前日期间的距离，从第 12 行的输出看，距离的精度是"天"。第 13 行输出了年龄"24 岁"。

第 20 行计算了两个时间对象之间的距离，从第 21 行的输出看，精度是纳秒。从第 23 行开始通过 toHours 等实例方法转换为以小时、分、秒、毫秒、纳秒为单位的距离。

源文件 6-25 展示了 ZonedDateTime 的用法。类方法 now 以形如 UTC+08:00 的时区标识来创建 ZonedDateTime 对象。"UTC+08:00"称为偏移量风格的时区标识。

源文件 6-25　ZonedDateTimeDemo.java

```
1 import java.time.ZonedDateTime;
2 import java.time.ZoneId;
3
4 public class ZonedDateTimeDemo {
```

```
5     public static void main(String[] args) {
6         System.out.println("Beijing current time: "
7             + ZonedDateTime.now(ZoneId.of("UTC+08:00")));
8         System.out.println("London current time: "
9             + ZonedDateTime.now(ZoneId.of("UTC+00:00")));
10    }
11 }
```

程序的输出为:

```
Beijing current time: 2024-01-26T05:55:28.885183100+08:00[UTC+08:00]
London current time: 2024-01-25T21:55:28.887164400Z[UTC]
```

6.7.3 时间戳

时间戳(Timestamp)指从一个预设的时间原点开始计数的经过时间单位数。这个原点通常是固定的,最常见的是自 1970 年 1 月 1 日 0 时 0 分 0 秒(UTC/GMT 时间)起算的秒数,也被称为 UNIX 时间戳或 Epoch 时间。时间戳具有全球一致性,不受时区影响。所以无论在地球上的哪个位置读取同一时间点产生的时间戳,其数值都是相同的。

日常生活中的日期和时间组,如"2024 年 2 月 5 日 星期一 14:30:00",包含了年、月、日、小时、分钟和秒等信息,并且依赖于具体的时区。而时间戳则是以统一标准格式表达的一个时间点,不直接包含时区信息,但可以转化为任何时区下的具体时间。时间戳是一个精确无歧义的数值,可以被计算机程序方便地存储和处理。Java 8 引入的 java.time.Instant 类是时间戳的一种实现。

给定飞机起飞时刻和落地时刻,源文件 6-26 中的程序计算飞行时长。第 7 行和第 10 行先定义了两个 Instant 对象分别引用起飞时刻和落地时刻。第 13 行通过 Duration.between()方法计算两者之间的距离,最后将飞行时长转换为小时和分钟进行输出。注意,如果实际应用中获取的是本地日期时间而非 UTC 时间,则应使用 LocalDateTime 配合 ZoneId 进行处理。

源文件 6-26 InstantDemo.java

```
1  import java.time.Duration;
2  import java.time.Instant;
3
4  public class InstantDemo {
5      public static void main(String[] args) {
6          //设起飞时刻为 2024-02-01T13:00:00Z(这是一个 UTC 时间)
7          Instant departure = Instant.parse("2024-01-31T13:00:00Z");
8
9          //设落地时刻为 2024-02-01T17:30:00Z
10         Instant arrival = Instant.parse("2024-02-01T17:30:00Z");
11
12         //计算飞行时长
13         Duration flightDuration = Duration.between(departure, arrival);
14
15         //输出飞行时长(以小时和分钟表示)
16         long hours = flightDuration.toHours();
17         int minutes = (int) (flightDuration.minusHours(hours).toMinutes());
18
```

```
19          System.out.println("飞行时长:");
20          System.out.println(hours + " 小时 " + minutes + " 分钟");
21      }
22 }
```

6.7.4 Date 类

java.util.Date 类表示系统特定的时间,可以精确到毫秒。Date 对象表示时间的默认顺序是星期、月、日、小时、分、秒、年。源文件 6-27 展示了 Date 的用法。

源文件 6-27　DateDemo.java

```
1  import java.util.Date;
2  import java.text.SimpleDateFormat;
3  import java.text.ParseException;
4
5  public class DateDemo {
6      public static void main(String[] args) throws ParseException{
7          //创建当前时间的 Date 对象
8          Date currentDate = new Date();
9          //通过毫秒值创建 Date 对象
10         long milliseconds = System.currentTimeMillis();
11         Date currentDateFromMillis = new Date(milliseconds);
12
13         //这种构造函数已被弃用,因为年份始于 1900,月份是从 0 开始的(0 表示一月)
14         //Date s = new Date(year, month, day, hour, minute, second);
15
16         //获取 Date 对象的信息
17         //getYear()返回的年份偏移了 1900,需要显式修正,已经被弃用
18         int year = currentDate.getYear() + 1900;
19         //getMonth()返回的月份从 0 开始,需要显式加 1 来修正,已经被弃用
20         int month = currentDate.getMonth() + 1;
21         //getDate()已经被弃用
22         int day = currentDate.getDate();
23         //getHours()已经被弃用
24         int hour = currentDate.getHours();
25         //getMinutes() 已经被弃用
26         int minute = currentDate.getMinutes();
27         //getSeconds() 已经被弃用
28         int second = currentDate.getSeconds();
29
30         //获取自 1970 年 1 月 1 日 0 点 0 分 0 秒(UTC)以来的毫秒数
31         long millisecond = currentDate.getTime();
32
33         //修改 Date 对象,这些 setXXX 方法均已经被弃用
34         currentDate.setYear(year - 1900);
35         currentDate.setMonth(month - 1);
36         currentDate.setDate(day);
37         currentDate.setHours(hour);
38         currentDate.setMinutes(minute);
39         currentDate.setSeconds(second);
40
41         //格式化和解析日期
```

```
42        SimpleDateFormat sdf = new SimpleDateFormat("yyyy-MM-dd HH:mm:ss");
43        String currentDateStr = sdf.format(currentDate);
44        Date currentDateFromString = sdf.parse(currentDateStr);
45    }
46 }
47 //代码来自通义千问,问题:Java Date 类的用法?
```

从 Java 8 开始,建议使用 java.time 包下的类替代 java.util.Date,因为它们解决了先前设计的问题,并提供了更好的 API 来处理日期和时间。

java.util.Date 类的大部分方法都已经废弃,并且在处理日期和时间时,易出现混淆和错误,尤其是时区问题。在新代码中尽量避免使用。

6.7.5 Calendar 类

java.util.Calendar 是一个抽象类,它提供了一种方法来处理日期和时间,支持时区转换。java.util.Calendar 的类方法 getInstance 用以创建 Calendar 对象,例如:

```
Calendartoday = Calendar. getInstance();
```

这里的 getInstance()方法返回一个 Calendar 实例,它被初始化为当前系统的日期和时间,并且使用系统默认的时区。然后可以通过各种重载的 set 方法修改日历对象的日期和时间:

```
public final void set (int year, int month, int day)
public final void set(int year, int month, int day, int hour, int minute)
public final void set (int year, int month, int day, int hour, int minute, int second)
```

参数 year 是负数时则表示公元前。

通过 Calendar 类中声明的常量,如 Calendar. YEAR,Calendar. MONTH,Calendar. DAY_OF_ MONTH,Calendar. DAY_OF_WEEK,Calendar. HOUR_ OF_DAY,Calendar. MINUTE,Calendar. SECOND 等,可以访问 Calendar 对象中的年、月、日、时、分、秒等分量。还提供了 add 等方法用于日期和时间的计算。源文件 6-28 演示 Calendar 类的使用。

源文件 6-28 CalendarDemo.java

```
1 import java.util.Calendar;
2 import java.util.Date;
3
4 public class CalendarDemo {
5     public static void main(String[] args) {
6         //获取当前日期和时间
7         Calendar calendar = Calendar.getInstance();
8         System.out.println(calendar);
9
10        //设置日期和时间
11        calendar.set(Calendar.YEAR, 2024);
12        //注意月份是从 0 开始的,所以 1 表示二月,以此类推
13        calendar.set(Calendar.MONTH, Calendar.JANUARY);
14        calendar.set(Calendar.DAY_OF_MONTH, 1);
15        calendar.set(Calendar.HOUR_OF_DAY, 10);
```

```
16          calendar.set(Calendar.MINUTE, 30);
17          calendar.set(Calendar.SECOND, 0);
18          calendar.set(Calendar.MILLISECOND, 0);
19          //转换为 Date 对象
20          System.out.println(calendar.getTime());
21
22          int year = calendar.get(Calendar.YEAR);
23          //为了得到正常的月份(1-12),需要加 1
24          int month = calendar.get(Calendar.MONTH) + 1;
25          int dayOfMonth = calendar.get(Calendar.DAY_OF_MONTH);
26
27          //加一天
28          calendar.add(Calendar.DAY_OF_MONTH, 1);
29
30          //减两小时
31          calendar.add(Calendar.HOUR_OF_DAY, -2);
32
33          //转换为 Date 对象
34          Date date = calendar.getTime();
35          System.out.println(date);
36      }
37 }
38 //代码来自通义千问,问题:Java Calendar 类的用法?
```

运行该程序,执行结果为:

```
java.util.GregorianCalendar[time = 1706222418902, areFieldsSet = true,
areAllFieldsSet= true, lenient = true, zone = sun.util.calendar.ZoneInfo[id=
"Asia/Shanghai", offset= 28800000, dstSavings= 0, useDaylight= false, transitions=
31, lastRule= null], firstDayOfWeek= 2, minimalDaysInFirstWeek= 1, ERA= 1, YEAR=
2024, MONTH= 0, WEEK_OF_YEAR= 4, WEEK_OF_MONTH= 4, DAY_OF_MONTH= 26, DAY_OF_YEAR= 26,
DAY_OF_WEEK= 6, DAY_OF_WEEK_IN_MONTH= 4, AM_PM= 0, HOUR= 6, HOUR_OF_DAY= 6, MINUTE=
40, SECOND= 18, MILLISECOND= 902, ZONE_OFFSET= 28800000, DST_OFFSET= 0]
Mon Jan 01 10:30:00 CST 2024
Tue Jan 02 08:30:00 CST 2024
```

其他方法还有:方法 clear()清零,getTimeInMillis()获取自 1970 年 1 月 1 日 0 点 0 分 0 秒(UTC)以来的毫秒数,setTimeZone(TimeZone zone)设置时区,getActualMaximum (int field)获取给定字段的最大可能值,例如某月的最大天数。

请注意,Java 8 及以后版本推荐使用 java.time 包中的 LocalDate、LocalTime、LocalDateTime、ZonedDateTime 等新类进行日期和时间操作,它们提供了更好的 API 设计和更高的精确度。然而对于兼容老版本或遗留代码,Calendar 类仍然有用。

6.8 Arrays 类

java.util.Arrays 类包含了对数组操作的各种方法,源文件 6-29 展示使用 Arrays 类的复制、排序、查找和填充等方法。

源文件 6-29　ArraysDemo.java

```java
1 import java.util.Arrays;
2
3 public class ArraysDemo {
4     public static void main(String[] args) {
5         int index = 0;
6         String key = new String();
7         String[] a = {"BJ", "TJ", "WH", "SH", "GZ", "SJZ"};
8         System.out.println(Arrays.toString(a));
9
10        //除了首尾元素把数组 a 复制到数组 b
11        String[] b;
12        b = Arrays.copyOfRange(a, 1, a.length - 1);
13        System.out.println(Arrays.toString(b));
14
15        //对数组排序
16        Arrays.sort(a);
17        System.out.println(Arrays.toString(a));
18
19        key = "SJZ";
20        index = Arrays.binarySearch(a, key);
21        System.out.println(key + "->" + index);
22        key = "CC";
23        index = Arrays.binarySearch(a, key);
24        System.out.println(key + "->" + index);
25
26        //除首尾元素外,其余填充为 NULL
27        Arrays.fill(a, 1, a.length - 1, "NULL");
28        System.out.println(Arrays.toString(a));
29    }
30 }
```

程序的输出为：

```
[BJ, TJ, WH, SH, GZ, SJZ]
[TJ, WH, SH, GZ]
[BJ, GZ, SH, SJZ, TJ, WH]
SJZ->3
CC->-2
[BJ, NULL, NULL, NULL, NULL, WH]
```

在源文件 6-29 中第 7 行创建了字符串数组,数组中是北京、天津、武汉、上海、广州和石家庄等城市的声母简拼。第 8 行使用类方法 toString 返回数组的字符串表示并输出。类方法 toString 有多种重载形式,使程序员避免通过循环语句逐个输出数组元素。

第 12 行使用类方法 copyOfRange 把数组 a 索引位置[1,5)中的元素复制到另外一个数组 b 中,数组 a 中共有 6 个元素,所以 a.length 的值为 6,那么 a.length－1 就是 5。

第 16 行对数组排序,排序后的数组就可以进行二分查找。第 20 行从数组中查找元素 SJZ,返回其索引位置 3。第 27 行使用 fill 方法按指定的元素和范围填充数组。

第 6 章　章节测验

第 7 章 JCF 框架

一组对象放在一起就成为一个对象集体(collection,a group of objects)。从对象间的前驱后继关系看,对象可以有三种组织管理方式:线性表(list)、树(tree)和图(graph)。如果约束每个对象只有唯一的前驱(除了第一个对象外)和唯一的后继(除了最后一个对象外),那么这种组织结构就是线性表。如果约束每个对象只有唯一前驱(除了一个称之为根的对象外),但可以有 0 个或若干个后继,那么这种组织结构就是树。如果每个对象允许有 0 个或若干个前驱,0 个或若干个后继,那么这种组织结构就是图。如果对象间没有前驱和后继关系的约束,但不允许有重复的对象,那么这种组织结构就是集合(set)。如果对象间没有前驱和后继关系的约束,还允许有重复的对象,那么这种组织结构就是集体。在集体上没有强制元素间的关系,可以认为没有结构。

JCF(Java Collection Framework)框架设计了用于存储和管理一组对象的接口和类以方便客户程序完成添加、删除、替换、查找和遍历等操作。

7.1　JCF 框架简介

JCF 框架由三部分组成:一组接口、一组实现类和一组算法。JCF 框架使用一组接口定义了实现类(功能提供者)与客户程序(功能使用者)之间的约定。这使得客户程序无须关心实现细节。其中有 4 个基本的接口:Collection<E>、Set<E>、List<E>和 Map<K,V>。其中 E、K、V 是类型形式参数,用来指定集体中元素的类型。这些接口的继承关系如图 7-1 所示。

Collection<E>接口是所有接口的祖先接口。尖括号括起来的 E 是类型参数。在程序中使用 Collection<E>时,需要使用具体的类型替换其中的 E。例如

```
Collection<String>  c;
```

就声明了一个引用类型的变量 c,这个变量引用元素为字符串的 Collection 对象,即引用一

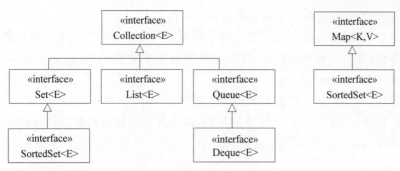

图 7-1 JCF 框架中的主要接口

组字符串。

接口 Collection<E> 类型的对象是容纳和操作一组对象的集体对象，集体中的元素没有前驱后继关系，并允许重复。

接口 Set<E> 定义了对数学中的集合对象进行元素的添加、删除等方法。接口 List<E> 定义了对线性表的一组操作，并且在线性表上定义了位置索引。线性表是元素有序的集体，并允许存放重复元素。可通过元素的位置索引直接访问线性表中的元素。接口 Set<E> 和 List<E> 都继承了接口 Collection<E>。队列是先进先出的线性表，队列接口 Queue<E> 应该是线性表接口的子接口，但在 Java SE 中设计为了 Collection<E> 的子接口。序集合接口 SortedSet 是 Set 的子接口。双端队列接口 Deque 是队列接口 Queue 的子接口。

接口 Map<K,V> 定义了<键,值>对(key-value pairs)的集合以及其上的各种操作方法。在<键,值>对中，每个键至多映射到一个值上。Map<K,V> 类型的集体对象不能包含重复的键。K 和 V 是类型参数，但必须是引用类型，即不能使用基本数据类型。Map<K,V> 类型的对象不是 Collection<E>，也不是 Set<E>。SortedMap 是 Map 的子接口。

理解了 JCF 定义的接口及其之间的关系，就理解了 JCF 的本质。JCF 还提供了一组标准的实现类实现这些接口。有一些实现类提供了完整的实现，客户程序可以直接使用。有一些实现类是抽象类，提供了各种实现类的共享功能。例如 Set<E> 的哈希表实现类 HashSet<E> 继承了 AbstractSet<E>。采用动态顺序表存储结构的线性表实现类 ArrayList<E> 继承了 AbstractList<E>。接口与实现的分离使得应用程序能够在不改变客户程序情况下，修改实现类。表 7-1 汇总了 JCF 中的接口与其常用的实现类。

表 7-1 JCF 中的接口与实现类

接 口	实 现 类	历 史
Set	HashSet,TreeSet,LinkedHashSet	
List	ArrayList,LinkedList	Vector,Stack
Queue	LinkedList,ArrayDeque,PriorityQueue	
Deque	ArrayDeque	
Map	HashMap,TreeMap,LinkedHashMap	Hashtable

JCF 还定义了一组算法实现用于进行查找、排序等操作。这些算法以静态方法的形式定义在 Collections 类中。注意，这个类名中有"s"。

设计 JCF 框架为应用程序开发提供了如下好处。

(1) 高性能,各种接口的实现类都选择了最优时间复杂性和空间复杂性的算法。

(2) 优雅性,在应用程序设计过程中,直接使用 JCF 提供的 API,而不必从头设计集体类的数据结构和算法。

(3) 一致性,不同类型的集体以类似的方法进行操作,比如排序。

(4) 可扩展性,只要实现了 JCF 中的接口,就允许使用新的数据结构和算法。

JDK 中对 Collection 接口的定义如下:

```java
public interface Collection<E> extends Iterable<E> {
    //基本操作
    //向集体中添加元素
    boolean add(E element);
    //从集体中删除参数指定的元素
    boolean remove(Object element);
    //判断集体中是否包含参数指定的元素
    boolean contains(Object element);
    //返回集体中实际元素的数目
    int size();
    //如果是空集体返回 true
    boolean isEmpty();
    //返回迭代器对象
    Iterator<E> iterator();

    //数组操作
    Object[] toArray();
    <T> T[] toArray(T[] a);
    ...
}
```

其中,add 方法用于向集体中添加指定类型的元素。如果添加成功,返回 true。当集体不允许重复元素,而此次添加会造成元素重复时,方法返回 false。其他任何添加失败的情况均抛出异常。比如,试图添加的对象无法转型到指定类型则抛出 ClassCastException。如果集体不允许 null 元素,而试图添加 null 元素,则抛出 NullPointerException 类型的异常。

remove 方法从集体中删除指定的元素。该方法会使用元素上的 equals 方法判断集体中哪个元素 e 是指定删除的对象 o:o.equals(e)。如果集体中含有多个重复元素,那么仅删除其中一个。

toArray 方法把当前的集体对象转换为数组。

对集体类型的对象进行遍历有两种方式:使用 for-each 循环,或者使用迭代器 Iterator。

for-each 语句是为遍历集体中或者数组中的元素而专门设计的一种简洁的语句。假设 collection 是 Collection<String> 类型的引用变量并已经引用了一个 Collection<String> 类型的对象,用 for-each 语句输出集体中的每个元素的语句为:

```
for (String str : collection){
    System.out.println(str);
}
```

这条语句读作：对 collection 中的每个字符串对象 str，使用 System.out 的 println 方法输出。语句中声明了引用类型的变量 str 用于逐个引用每个元素。

迭代器在封装了 collection 的存储结构和存储设施的前提下提供了遍历 collection 的功能。也就是说，客户程序不必关心 collection 的存储结构就能够遍历 collection。迭代器对象可由 Collection 对象的实例方法 iterator 获取，该方法返回 Iterator 类型的对象。下面是 Java 中 Iterator 接口的定义：

```
public interface Iterator<E> {
    //如果集体中还有元素返回 true
    boolean hasNext();
    //返回下个元素
    E next();
    ……
}
```

其中，hasNext 方法用于查看迭代器中是否还有元素没有访问；next 方法返回本次迭代中的下个元素。有了用于遍历集体对象的迭代器对象后，就用该对象上的 next 方法获取集体中的下个对象。hasNext 方法判断 collection 中是否还有元素。

假设 collection 是 Collection<String>类型的引用变量并已经引用了 Collection<String>对象。下面的代码片段使用迭代器 iter 遍历 collection：

```
Iterator<String>  iterator = collection.iterator();
while (iterator.hasNext()) {
    System.out.print(iterator.next());
}
```

JCF 把一些常规性的算法以类方法的形式声明在类 Collections 中。这些算法包括求最大元素、最小元素、查询元素的出现频数、判断两个 Collection 对象是否相交等。表 7-2 是 Collections 类中的部分类方法。详细的类方法请查阅 The Java Platform，Standard Edition（Java SE）APIs 中 java.base 模块 java.util 包的帮助文档（API documentation）。

表 7-2 Collections 类中的部分类方法

方法签名	功　　能
disjoint(Collection c1,Collection c2)	如果两个 Collection 对象 c1 和 c2 没有共同的元素则返回 true
frequency(Collection c,Object o)	返回对象 o 在 Collection 集体 c 中出现的频数
max(Collection c)	返回 Collection 类型集体中最大的元素。使用根据元素上定义的 compareTo 方法比较元素的大小
min(Collection c)	返回 Collection 类型集体中最小的元素。使用根据元素上定义的 compareTo 方法比较元素的大小

7.2 Set 接口和实现类

Set 类型是数学中"集合"概念在 Java 中的实现。在 Set 类型的集体中不允许有重复元素。也就是说,假设 e1 和 e2 都是 E 类型的对象,如果 e1.equals(e2),那么 e1 和 e2 就不能同时放在集体中。另外,Set 类型的集体至多允许有一个 null 元素。

下面是 Java SE 中对 Set 接口的定义:

```java
public interface Set<E> extends Collection<E> {
    //添加元素
    boolean add(E e);
    //删除元素
    boolean remove(Object o);
    //是否包含参数指定的元素
    boolean contains(Object element);
    int size();
    boolean isEmpty();
    void clear();

    Iterator<E> iterator();

    //数组操作
    Object[] toArray();
    <T> T[] toArray(T[] a);
}
```

Set 接口首先继承了 Collection 接口。这样就继承了 add 方法、remove 方法、contains 方法等。在 add 方法中,如果指定的参数 e 没有在集合中,那么就把参数 e 引用的对象添加到集合中并返回 true;否则返回 false,即对象未能成功加入集合中。

HashSet 是 Set 接口的实现类之一。HashSet 使用哈希表作为存储结构。如果想使用 HashSet 实现类,那么作为元素的对象必须覆盖 hashCode 方法和 equals 方法。因为 HashSet 类使用 equals 方法判断两个元素是否重复,使用 hashCode 方法计算哈希表使用的哈希地址。

源文件 7-1　**SetDemo.java**

```java
1  import java.util.Set;
2  import java.util.HashSet;
3  import java.util.Iterator;
4
5  public class SetDemo {
6      public static void main(final String[] args) {
7          Set<String> h = new HashSet<>();
8          System.out.println("Is the set empty? " + h.isEmpty());
9          h.add("A");
10         h.add("B");
11         h.add("C");
12         h.add("D");
13         h.add("E");
```

```
14          h.add(null);
15          h.add("B");
16          h.add(null);
17          System.out.println("The number of elements in the set: "
18              + h.size());
19          System.out.println(h);
20
21          System.out.println("Does the set contain null? "
22              + h.contains(null));
23
24          //使用对象的 equals 方法判断两个元素是否相等.
25          //如果找到与参数相等的元素,则删除并返回 true;否则返回 false
26          System.out.println("Is null removed? " + h.remove(null));
27          System.out.println(h);
28
29          //通过迭代器遍历
30          Iterator<String> iter = h.iterator();
31          while (iter.hasNext()) {
32              System.out.print(iter.next());
33          }
34          System.out.println();
35
36          //通过 for each 循环遍历
37          for (String e : h) {
38              System.out.print(e);
39          }
40          System.out.println();
41
42          //清空
43          h.clear();
44          System.out.println("Is the set cleared? " + h.isEmpty());
45      }
46  }
```

在源文件 7-1 中,第 7 行首先创建了 HashSet<String>类型的集合对象。类型参数 String 声明了集合元素的类型为 String。这个类型参数是编译器从引用变量 h 的类型 Set <String>中推导来的:如果引用变量的集体类型中每个元素是 String 类型,那么集体对象中每个元素也是 String 类型。所以第 7 行使用 HashSet<>()而没有使用 HashSet<String>(),这使得源代码更加简洁。初始时集合是空的,所以第 8 行的 isEmpty 方法返回 true。然后使用 add 方法向集合中依次添加仅含一个字符的字符串 A,B,C,D,E。接着使用 add 方法添加 null,然后继续添加 B。由于 B 已经在集合中,所以添加失败,add 方法返回 false。然后继续添加 null,由于 null 已经在集合中,所以添加失败,add 方法返回 false。一系列添加元素的语句后,第 17 行使用 size 方法返回集合中当前元素的个数。然后程序输出集合,即输出集合中的每个元素。截至此处,程序的输出应是:

```
Is the set empty? true
The number of elements in the set: 6
[null, A, B, C, D, E]
```

然后第 21 行使用 contains 方法判断集合中是否有 null,方法的返回值是 true。接着程序使用 remove 方法从集合中删除 null。此时再次输出集合中的所有元素。截至此处,程序的输出应是：

```
Does the set contain null? true
Is null removed? true
[A, B, C, D, E]
```

注意：System.out.println(h)实际上是 System.out.println(h.toString())。

HashSet 继承了父类 AbstractCollection 的 toString 方法，该方法返回 Collection 对象的字符串表示。Collection 对象中的元素使用 String.valueOf(Object)方法转换成字符串。如果有多个元素，则使用逗号隔开。整个 Collection 对象的字符串表示括在方括号中。

接下来程序使用迭代器对象遍历整个集合。首先第 30 行使用 iterator 方法获取迭代器对象，然后使用迭代器对象上的 next 方法逐一输出每个元素。输出是：

```
ABCDE
```

接下来第 37 行使用 for-each 语句输出所有元素：

```
DEABC
```

最后程序使用 clear 方法清除所有元素，使得集合为空。所以此时使用 isEmpty 方法的返回结果为 true。

这里例子还演示了 null 也是集合中的一个元素，但是至多有一个。完整的程序输出为：

```
Is the set empty? true
The number of elements in the set: 6
[null, A, B, C, D, E]
Does the set contain A? true
Is null removed? true
[A, B, C, D, E]
ABCDE
ABCDE
Is the set cleared? true
```

当调用 add 方法添加元素时，如果该元素已经在集合中存在，那么添加失败返回 false。add 方法需要把试图添加的元素与集合中当前的所有元素比较，这就需要调用元素上的 equals 方法。String 类中已经声明了 equals 方法，而当使用自定义类型的对象作为集合元素，而不是使用 Java SE 中的类型，就需要显式给出 equals 方法的定义。

HashSet 类型的集合使用哈希表作为存储结构，默认的哈希表的容量初始值为 16(2 的 4 次方)。如果 16 个存储单元中 12 个(75%)有数据的时候，则认为加载因子达到默认值 0.75。这时 HashSet 抛弃当前的哈希表，重新开辟一个容量约为 32(2 的 5 次方)的哈希表，并重新计算安排各个对象。当把一个新的元素添加到哈希表的时候，需要根据该元素计算哈希地址。对象的 hashCode 方法就是把一个对象映射到一个整数用来计算哈希地址。所以，添加的对象如果是自定义类型，还必须实现 hashCode 方法。

源文件 7-2 中声明了自定义类 Car，其中有两个属性：车牌号码 licenceNumber 和质量 weight。第 18 行定义了含参数的构造方法。第 28 行重载了 hashCode 方法；第 41 行重载了 equals 方法。

源文件 7-2   CarSetDemo.java

```java
1 import java.util.HashSet;
2 import java.util.Set;
3
4 public class CarSetDemo {
5     public static void main(String[] args) {
6         Set<Car> s = new HashSet<>();
7         s.add(new Car("1234", 16000));
8         if (! s.add(new Car("1234", 16000))) {
9             System.out.println("Failed to add duplicated elements.");
10        }
11    }
12 }
13
14 class Car {
15     private String licenceNumber;
16     private int weight;
17
18     Car(String licenceNumber, int weight) {
19         super();
20         this.licenceNumber = licenceNumber;
21         this.weight = weight;
22     }
23
24     /*
25      * @see java.lang.Object#hashCode()
26      */
27     @Override
28     public int hashCode() {
29         final int prime = 31;
30         int result = 1;
31         result = prime * result
32                 + ((licenceNumber == null) ? 0 : licenceNumber.hashCode());
33         result = prime * result + weight;
34         return result;
35     }
36
37     /*
38      * @see java.lang.Object#equals(java.lang.Object)
39      */
40     @Override
41     public boolean equals(Object obj) {
42         if (this == obj) {
43             return true;
44         }
45
46         if (obj == null) {
47             return false;
48         }
49
50         if (getClass() != obj.getClass()) {
51             return false;
52         }
```

```
53          Car other = (Car) obj;
54          if (licenceNumber == null) {
55              if (other.licenceNumber != null) {
56                  return false;
57              }
58          } else if (! licenceNumber.equals(other.licenceNumber)) {
59              return false;
60          }
61          if (weight != other.weight) {
62              return false;
63          }
64          return true;
65      }
66  }
```

在第 5 行开始的 main 方法中,首先添加了一个车牌号码为"1234"、重量为 16000 的 Car 类型对象后,然后试图添加同样的对象,那么添加失败,程序输出 Failed to add duplicated elements。这是因为在 add 方法中会调用元素 Car 对象上的 equals 方法判断是否是重复的对象。

注意:HashSet 的存储是无序的,没有前后关系。hashCode 方法必须和 equals 方法一致:equals 方法判断相等的对象的 hashCode 方法返回值也相等;反过来,hashCode 方法返回值相等的对象的 equals 方法返回值未必相等。

Set 接口从 Collection 接口继承了批量操作方法,使用这些批量操作方法能够完成集合的并、交、差等运算。假设 A 和 B 是两个集合对象。集合 A 和集合 B 并集(A∪B)中的元素或者属于集合 A 或者属于集合 B。集合 A 和集合 B 的交集(A∩B)是既属于集合 A 又属于集合 B 中的元素的集合。A 和 B 的差集(A－B)指属于集合 A 但不属于集合 B 的元素的集合。

求 A∪B:A.addAll(B)。

求 A∩B:A.retainAll(B)。

求 A－B:A.removeAll(B)。

B 是 A 的子集? A.containsAll(B)。

不同于数学记号,这里的 A 和 B 都是集合 Set 类型的引用变量。为了计算集合 A 和 B 的并、交和差,必须创建新的集合对象存放结果。下面的代码片段使用引用变量 C 引用存放结果的集合:

```
//C = A ∪ B
Set<E> C = new HashSet<E>(A);
C.addAll(B);

//C = A ∩ B
Set<E> C = new HashSet<E>(A);
C.retainAll(B);

//C = A - B
Set<E> C = new HashSet<E>(A);
C.removeAll(B);
```

可以看到,以上完成集合并、交和差运算的代码片段的共同特点是,先把集合 A 添加到集合 C 中,然后使用集合 B 对集合 A 进行调整。源文件 7-3 演示了集合的并、交和差运算。

源文件 7-3　SetOperatorDemo.java

```java
1  import java.util.Set;
2  import java.util.HashSet;
3
4  public class SetOperatorDemo {
5      public static void main(String[] args) {
6          String[] a = {"A", "B", "C", "D", "E"};
7          String[] b = {"X", "Y", "E"};
8          Set<String> aSet = new HashSet<>();
9          Set<String> bSet = new HashSet<>();
10
11         for (String e : a) {
12             aSet.add(e);
13         }
14
15         for (String e : b) {
16             bSet.add(e);
17         }
18
19         Set<String> cSet;
20         cSet = new HashSet<>(aSet);
21         //并
22         cSet.addAll(bSet);
23         System.out.println(cSet);
24
25         //交
26         cSet = new HashSet<>(aSet);
27         cSet.retainAll(bSet);
28         System.out.println(cSet);
29
30         //差
31         cSet = new HashSet<>(aSet);
32         cSet.removeAll(bSet);
33         System.out.println(cSet);
34     }
35
36 }
```

第 6 行和第 7 行首先创建了两个字符串数组，第 8 行和第 18 行使用这两个字符串数组初始化了两个集合对象 aSet 和 bSet。集合 aSet 初始元素是 A、B、C、D、E，集合 bSet 的初始元素是 X、Y、E。第 20 行以集合 aSet 作为参数构造集合对象 cSet，第 22 行与集合 bSet 并运算的结果是：

[A, B, C, D, E, X, Y]

第 26 行以集合 aSet 作为参数构造集合对象 cSet，第 27 行集合 cSet 和 bSet 交运算的结果是：

[E]

第 31 行以集合 aSet 作为参数构造集合对象 cSet，第 32 行 cSet 与集合 bSet 差运算的结果是：

[A, B, C, D]

完整的程序输出为：

```
[A, B, C, D, E, X, Y]
[E]
[A, B, C, D]
```

7.3 List 接口

List 接口是对线性表数据结构的抽象。在线性表中,除了第一个元素外,每个元素都有唯一前驱;除了最后一个元素外,每个元素都有唯一后继。因此,线性表中的元素就有了位置索引。线性表在继承了 Collection 接口的基础之上,还提供了按照位置索引对元素进行增加、删除、替换、查找等操作。一般把这些线性表所特有的操作分成四类:基于位置索引的增、删、查、改;查找指定的对象,返回其在线性表中的位置索引;遍历和截取任意子表。

List 接口的定义如下所示。

```java
public interface List<E> extends Collection<E> {
    //基本操作
    //添加元素
    boolean add(E element)
    //在参数指定索引位置前插入元素
    void add(int index, E element)
    //删除参数指定的元素
    boolean remove(Object o)
    //删除参数指定索引位置后的元素
    E remove(int index)
    //读取参数指定索引位置后的元素
    E get(int index);
    //替换参数指定索引位置后的元素
    E set(int index, E element)
    int size();
    boolean isEmpty();

    //查找
    int indexOf(Object o);
    int lastIndexOf(Object o);
    //如果线性表中含有参数指定元素则返回 true
    boolean contains(Object o)

    //迭代器
    ListIterator<E> listIterator();
    ListIterator<E> listIterator(int index);

    //数组操作
    //转换为数组
    Object[] toArray()
}
```

List 接口中的 add 方法覆盖了 Collection 接口中的 add 方法,声明抛出同样的异常。但与 Set 接口中的 add 方法不同:把指定的对象添加到线性表的末尾,并且允许重复。List

接口还提供了特有的基于位置索引的添加方法：add(int index，E element)用于在指定的位置索引前添加对象。

remove(Object o)删除对象 o 在线性表中的第一次出现。remove(int index)删除指定位置索引 index 后的对象。size 方法返回当前元素的个数。isEmpty 方法判断当前线性表是否为空。

get(int index)返回指定位置索引 index 后的对象。set(int index，E element)设置指定位置索引 index 后的对象为 element。contains(Object o)判断线性表中是否包含对象 o。indexOf(Object o)返回指定对象 o 在线性表中第一次出现的位置索引，lastIndexOf(Object o)返回指定对象 o 在线性表中最后一次出现的位置索引。如果没有找到，则返回一1。

通过 listIterator 方法获取的迭代器对象不仅支持 hasNext 和 next 方法，还支持 hasPrevious 和 previous 方法，用于访问前一个元素。

subList(int from，int to)方法按照指定的范围返回一个局部的线性表。

JCF 中有两个 List 接口的实现类：ArrayList 和 LinkedList。前者使用动态顺序存储结构；后者使用链表作为存储结构。

7.3.1　ArrayList 实现类

ArrayList 使用容量动态可变的数组作为线性表的存储结构。每个 ArrayList 对象都有一个容量属性(capacity)，当添加元素超过容量限制时，容量会自动增长，其默认初值为 10。

图 7-2 展示了接口 Collection、List 和类 ArrayList 之间的关系：List 继承了 Collection；ArrayList 实现了 List。

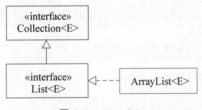

图 7-2　ArrayList

List 类型的集体比 Collection 类型的集体多了"位置索引"属性。在一个长度为 n 的 List 集体中，有 n+1 个有效的位置索引：从 0 到 n，包括 0 和 n。位置索引并不在元素之上，而是在两个元素之间。如图 7-3 所示。其中的 List 集体中有 4 个元素，5 个位置索引：0，1，2，3，4。

图 7-3　ArrayList 的位置索引

迭代器的 previous 方法返回位置索引 i 之前的元素；next 方法则返回位置索引之后的元素。

源文件 7-4 演示了 ArrayList 类的用法。包括 List 类型集体对象的创建，元素的添加、删除等。

源文件 7-4　ArrayListDemo.java

```java
1  import java.util.List;
2  import java.util.ListIterator;
3  import java.util.ArrayList;
4
5  public class ArrayListDemo {
6      public static void main(String[] args) {
7          String[] items = {"A", "B", "C", "D", "E", null, "F"};
8          List<String> a = new ArrayList<>();
9          for (String item : items) {
10             a.add(item);
11         }
12         System.out.println(a);
13
14         System.out.println("The element at 5: " + a.get(5));
15         System.out.println("Replace the element at 5 with \'X\': "
16                 + a.set(5, "X"));
17         System.out.println(a);
18
19         //插入元素
20         System.out.println("Insert \'Y\' before 6!");
21         a.add(6, "Y");
22         System.out.println(a);
23
24         System.out.println("Remove the element at 3: " + a.remove(3));
25         System.out.println(a);
26
27         //向后遍历
28         ListIterator<String> iter = a.listIterator();
29         while (iter.hasNext()) {
30             System.out.print(iter.next());
31         }
32         System.out.println();
33
34         //向前遍历
35         //a.size() 返回实际元素个数,以此数为迭代器的初始位置的索引值
36         ListIterator<String> it = a.listIterator(a.size());
37         while (it.hasPrevious()) {
38             System.out.print(it.previous());
39         }
40     }
41 }
```

程序的输出为：

```
[A, B, C, D, E, null, F]
The element at 5: null
Replace the element at 5 with 'X': null
[A, B, C, D, E, X, F]
Insert 'Y' before 6!
[A, B, C, D, E, X, Y, F]
Remove the element at 3: D
```

```
[A, B, C, E, X, Y, F]
ABCEXYF
FYXECBA
```

为了创建一个 List 对象并把初始的元素设置为字符串对象"A"、"B"、"C"、"D"、"E"、null、"F",源文件 7-4 首先在第 8 行使用这些字符串对象构造了一个字符串数组 items。然后使用 for-each 循环把数组 items 中每个元素添加到 List 对象中。这样,该线性表的初始元素就是这 6 个字符串对象以及 null,共 7 个对象,并通过 List<String> 类型的变量 a 引用。

在第 12 行以默认格式输出线性表 a 中的所有对象:

```
[A, B, C, D, E, null, F]
```

此时的位置索引应是:

$_0A_1B_2C_3D_4E_5null_6F_7$

因此,第 14 行使用 get 方法获取位置索引 5 之后的元素即是 null:

```
The element at 5: null
```

第 16 行使用 set 方法把位置索引 5 之后的元素改为字符串对象"X",然后在第 17 行输出,确认把 null 更改为了"X":

```
[A, B, C, D, E, X, F]
```

那么,当前线性表 a 中索引情况是:

$_0A_1B_2C_3D_4E_5X_6F_7$

接着程序在第 17 行使用 insert 方法把字符串对象"Y"插入在位置索引 6 之前,然后在第 18 行再次输出线性表 a,结果显示字符串对象"Y"插入到了位置索引 6 之前:

```
[A, B, C, D, E, X, Y, F]
```

第 24 行使用 remove 方法删除位置索引 3 之后的元素,然后输出,结果显示位置索引 3 之后的字符串对象"D"被删除了:

```
[A, B, C, E, X, Y, F]
```

从第 28 行到第 32 行使用迭代器遍历线性表 a 中的所有对象。首先使用 listIterator 方法获取线性表 a 的迭代器,默认从前到后进行迭代,然后使用 next 方法获取下元素。结果为:

```
ABCEXYF
```

程序从第 36 行到第 39 行使用迭代其逆向遍历线性表对象 a 中的全部对象。当前线性表 a 中的元素及索引情况是:

$_0A_1B_2C_3E_4X_5Y_6F_7$

listIterator 方法允许以参数指定迭代起始位置。第 36 行试图从最后一个位置索引开始迭代,所以需要获取线性表 a 的最后一个位置索引。线性表的最后一个位置索引刚好和线性表当前元素的个数相等,所以该位置索引可以通过线性表对象 a 的 size 方法获取。第

36 行的 a.size() 返回 7,那么 a.listIterator(7)就从线性表对象 a 上返回从位置索引 7 开始迭代的迭代器 it。

A B C E X Y F₇

迭代器的初始位置就是 7。在 7 这个位置索引上,it.previous()返回字符串"F"并使迭代器的当前位置向前移动。

A B C E X Y₆ F

在这个位置索引,it.previous()返回字符串对象"Y",并使迭代器的当前位置前移：

A B C E X₅ Y F

在这个位置索引,it.previous()返回字符串"X",并使迭代器的当前位置前移。当 it.previous()返回字符串对象"A"后,迭代器的情形是：

0A B C E X Y F

在这个位置索引,it.hasPrevious()返回 false 从而结束循环。

注意：一般使用接口 List 声明变量,引用具体实现类创建的对象,就像第 8 行那样；如果 List 线性表中有 n 个元素,那么有 n+1 个位置索引：从 0 到 n；insert 方法把对象插入在指定的位置索引之前；其他方法,如 remove 方法,都是操作指定位置索引后面的元素；使用 ListIterator 迭代器既可以自前向后迭代,也可以自后向前迭代；仅能向 Collection 类型的线性表中添加对象,而不能添加基本数据类型的数据。

7.3.2 LinkedList 实现类

JDK 中的 LinkedList 类使用双向链表作为存储结构实现 List 接口。LinkedList<E>内部以链表来保存线性表中的元素,因此随机访问元素时的性能较差,但在插入、删除元素时性能较好。ArrayList<E>内部以数组存储线性表中的元素,因此随机访问元素时的性能较好,但在插入、删除元素时性能较差。

当线性表中存在重复元素 o 时,可使用实例方法 indexOf 查找线性表中 o 的第一次出现位置的索引。如果在线性表中找不到指定的元素 o,返回－1。使用实例方法 lastIndexOf 查找线性表中指定元素 o 的最后一次出现位置的索引。如果在线性表中找不到元素 o,返回－1。源文件 7-5 展示了 LinkedList<E>的用法。在第 6 行创建了元素为 Integer 类型的 LinkedList 线性表对象,从第 8 行到第 13 行向线性表中添加 1、2、3、4、2、3,这些整数字面量会被自动装箱为 Integer 对象后添加到线性表中。

源文件 7-5　LinkedListExample.java

```
1 import java.util.List;
2 import java.util.LinkedList;
3
4 public class LinkedListExample {
5     public static void main(String[] args) {
6         List<Integer> linkedListNumbers = new LinkedList<>();
7
8         linkedListNumbers.add(1);
9         linkedListNumbers.add(2);
```

```
10          linkedListNumbers.add(3);
11          linkedListNumbers.add(4);
12          linkedListNumbers.add(2);
13          linkedListNumbers.add(3);
14
15          int index = linkedListNumbers.indexOf(3);
16          if (index == -1) {
17              System.out.println("Element not found");
18          } else {
19              System.out.println("Element found at index " + index);
20              linkedListNumbers.remove(index);
21          }
22          System.out.println(linkedListNumbers);
23
24          index = linkedListNumbers.lastIndexOf(3);
25          if (index == -1) {
26              System.out.println("Element not found");
27          } else {
28              System.out.println("Last occurrence at index " + index);
29              linkedListNumbers.add(index, 5);
30          }
31          System.out.println(linkedListNumbers);
32
33          System.out.println(linkedListNumbers.contains(5));
34          System.out.println(linkedListNumbers.indexOf(5));
35      }
36 }
```

第 15 行到第 22 行查找元素 3 首次出现的位置,如果找到了,输出其位置索引并将其删除。程序将输出:

```
Element found at index 2
[1, 2, 4, 2, 3]
```

第 24 行到第 30 行查找元素 3 最后一次出现的位置,如果找到了,输出其位置索引并在该位置前插入 5。插入 5 后的线性表为:

```
[1, 2, 4, 2, 5, 3]
```

第 33 行使用 contains 方法判断线性表中是否包含元素 5;第 34 行返回元素 5 的首次出现位置的索引。

完整的程序输出为:

```
Element found at index 2
[1, 2, 4, 2, 3]
Last occurrence at index 4
[1, 2, 4, 2, 5, 3]
true
4
```

7.3.3 Collections 类

以线性表组织管理一组对象,除了前文介绍的增删查改等面向单个元素的操作外,还涉及一些面向线性表的整体操作，如排序等。Java 把这些面向整个线性表的操作的实现算法以静态方法形式安排在 Collections 类中。这些方法的参数都是 List 类型。具体如表 7-3 所示。

表 7-3　Collections 中的静态方法

方　法　名	功　　能
sort	使用归并排序算法排序
binarySearch	使用二分查找算法查找指定的对象。找到则返回其位置索引，否则返回－1
copy	复制
fill	使用指定的对象替换所有元素，相当于填充
reverse	反转
rotate	按照指定的距离（distance）旋转
swap	交换指定位置索引的两个元素
shuffle	随机排列元素
replaceAll	使用另外一个对象替换指定对象的全部出现
indexOfSubList	在一个子表中查找指定对象的首次出现。找到则返回其位置索引，否则返回－1

　　类 Collections 的静态方法 sort 对指定 List 类型线性表中所有对象进行排序。无论使用什么样的排序算法，总得比较元素大小。如果把基本数据类型的数据，比如 int 类型的数据安排在数组中，然后对这个数组进行排序，可以使用比较运算符"＞"完成两个元素的比较。现在的问题是，List＜E＞类型的线性表中存放的是 E 类型的对象。对于两个对象进行大小比较，比如比较两个字符串对象，就不能使用关系运算符"＞"。事实上，sort 方法默认去调用元素上的 compareTo 方法进行比较。

　　假设变量 a 和 b 引用了两个字符串对象，那么就是使用 a.compareTo(b) 进行比较：如果 a 小于 b，那么 compareTo 方法返回一个负数；如果 a 等于 b，那么 compareTo 方法返回 0；如果 a 大于 b，那么 compareTo 返回一个正数。

　　Comparable 接口是 Collections 的静态方法 sort（客户程序）与被排序线性表间的一个约定：约定使用 compareTo 方法比较线性表中的元素。如果元素的类型是 Comparable 的实现类，那么 sort 方法就能调用其 compareTo 方法；如果元素的类型不是 Comparable 的实现类，那么 sort 方法就不能进行排序。

　　源文件 7-6 演示了如何使用 sort 方法对元素类型 String 的 List 线性表进行排序。由于 String 类实现了 Comparable 接口，这种情况下 sort 方法调用 String 类的 compareTo 方法进行元素比较。

源文件 7-6　CollectionsSortDemo.java

```
1  import java.util.Collections;
2  import java.util.ArrayList;
3  import java.util.List;
4
5  public class CollectionsSortDemo {
6      public static void main(String[] args) {
7          List<String> list = new ArrayList<>();
```

```
8          list.add("apple");
9          list.add("app");
10         list.add("bed");
11         System.out.println(list);
12         Collections.sort(list);
13         System.out.println(list);
14     }
15 }
```

程序的输出为：

```
[apple, app, bed]
[app, apple, bed]
```

String 类的 compareTo 方法按照字符串中每个字符的 Unicode 代码点依次进行大小比较。假设变量 a 和 b 引用两个字符串对象，那么当 a 和 b 两个字符串中每个字符对应相等时，a.compareTo(b) 返回 0；当 a 中相等的前缀后的某个字符小于 b 中对应的字符时，a.compareTo(b) 返回一个负数；当 a 中相等的前缀后的某个字符大于 b 中对应的字符时，a.compareTo(b) 返回一个正数；如果 a.compareTo(b) 返回 0，那么 a.equals(b) 返回 true；反过来，如果 a.equals(b) 返回 true，那么 a.compareTo(b) 返回 0。

如果线性表中的元素是 LocalDate 类型，那么就按照日期的前后进行大小比较，越早的日期越小。由于 Java 已经让 String 类和 LocalDate 类实现了 Comparable 接口，所以对 String 类型和 LocalDate 类型的对象就可以直接使用 Collections.sort 方法进行排序。

Comparable 接口约定了对象间的自然顺序（natural ordering），Collections.sort 方法自动地调用对象上的 compareTo 方法进行大小比较。如果想让自定义类的对象也能够使用 Collections.sort 方法进行自动排序，则该类必须实现 Comparable 接口。除此以外，还必须重载 equals 方法和 hashCode 方法以确保其与 JCF 框架一致的行为。

注意：虽然 java.lang.Object 类提供了默认的 equals 方法和 hashCode 方法的实现，但这些实现并不适合自定义类。

在 Comparable 接口中仅声明了一个抽象方法：

```
public interface Comparable<T> {
    public int compareTo(T o);
}
```

compareTo 方法把当前对象（this）与通过参数指定的对象 o 进行比较。当前对象小于、等于或者大于对象 o 时分别返回负整数、0 和正整数。源文件 7-7 演示了如何对自定义类的对象进行排序。有三个步骤：①自定义类并实现 Comparable 接口；②把自定义类的若干对象添加到线性表中；③使用 Collections.sort 对线性表排序。

源文件 7-7　CollectionsSortExample.java

```
1 import java.util.Collections;
2 import java.util.List;
3 import java.util.ArrayList;
4
5 class Car implements Comparable<Car> {
6     private String licenceNumber;
7     private Integer weight;
```

```
8
9       Car(String licenceNumber, int weight) {
10          this.licenceNumber = licenceNumber;
11          this.weight = weight;
12      }
13
14      public int compareTo(Car t) {
15          return this.weight.compareTo(t.weight);
16      }
17
18      @Override
19      public String toString() {
20          return "Car [" + licenceNumber + ", " + weight + "]";
21      }
22  }
23
24  public class CollectionsSortExample {
25      public static void main(String[] args) {
26          Car[] a = {new Car("B12", 18), new Car("A12", 20),
27                     new Car("Z34", 17), new Car("Z34W", 18),
28                     new Car("12NN", 19), new Car("12N", 20) };
29          List<Car> cars = new ArrayList<>();
30          for (Car c : a) {
31              cars.add(c);
32          }
33          System.out.println(cars);
34          Collections.sort(cars);
35          System.out.println(cars);
36      }
37  }
```

在源文件 7-7 中第 5 行轿车类 Car 被声明实现 Comparable 接口，在类的定义中提供了接口中约定的抽象方法 compareTo 的实现。该实现很简单，仅使用轿车的属性"重量"进行 Car 对象的大小比较。也就是说，有两个 Car 对象 a 和 b，a 的重量大于 b 的重量，那么对象 a 就大于对象 b。对于重量的比较使用了包装类 Integer 上定义的 compareTo 方法。

然后程序第 26 行创建了 6 个 Car 类型的对象，第 34 行使用 Collections.sort 方法进行排序。Collections.sort 方法会调用 Car 类型对象上的 compareTo 方法执行排序进行对象比较。

程序的输出为：

```
[Car [B12, 18], Car [A12, 20], Car [Z34, 17], Car [Z34W, 18], Car [12NN, 19], Car [12N, 20]]
[Car [Z34, 17], Car [B12, 18], Car [Z34W, 18], Car [12NN, 19], Car [A12, 20], Car [12N, 20]]
```

由于 compareTo 方法认为对象的 weight 属性相等对象就相等，所以还得定义 Car 类的 equals 方法和 hashCode 方法以保持一致性。这是因为 compareTo 方法返回 0 当且仅当 equals 方法返回 true；如果 equals 方法返回 true，hashCode 方法的返回值相等。

源文件 7-8 在源文件 7-7 的基础之上增加了查询最大元素和二分查找某个元素的功能，查找功能需要调用元素上的 equals 方法。

源文件 7-8　CollectionsSortSearchExample.java

```java
1  import java.util.Collections;
2  import java.util.List;
3  import java.util.ArrayList;
4
5  class Car implements Comparable<Car> {
6      private String licenceNumber;
7      private Integer weight;
8
9      Car(String licenceNumber, int weight) {
10         this.licenceNumber = licenceNumber;
11         this.weight = weight;
12     }
13
14     public int compareTo(Car t) {
15         return this.weight.compareTo(t.weight);
16     }
17
18     public int hashCode() {
19         final int prime = 31;
20         int result = 1;
21         result = prime * result
22             + ((weight == null) ? 0 :  weight.hashCode());
23         return result;
24     }
25
26     public boolean equals(Object obj) {
27         if (this == obj)
28             return true;
29         if (obj == null)
30             return false;
31         if (getClass() != obj.getClass())
32             return false;
33         Car other = (Car) obj;
34         if (weight == null) {
35             if (other.weight != null) {
36                 return false;
37             }
38         } else { if (! weight.equals(other.weight)) {
39             return false;
40             }
41         }
42         return true;
43     }
44
45     @Override
46     public String toString() {
47         return "Car [" + licenceNumber + ", " + weight + "]";
48     }
49 }
50
51 public class CollectionsSortSearchExample {
52     public static void main(String[] args) {
```

```
53        Car[] a = {new Car("B12", 18), new Car("A12", 20),
54                    new Car("Z34", 17), new Car("Z34W", 18),
55                    new Car("12NN", 19), new Car("12N", 20) };
56        List<Car> cars = new ArrayList<>();
57        for (Car c : a) {
58            cars.add(c);
59        }
60        System.out.println(cars);
61        Collections.sort(cars);
62        System.out.println(cars);
63        System.out.println(Collections.max(cars));
64        int position = Collections.binarySearch(cars,
65            new Car("A12", 20));
66        System.out.println("Position of the searched car = "
67            + position);
68    }
69 }
```

第 64 行 Collections.binarySearch 方法会调用第 26 行定义的 equals 方法比较 Car 对象。程序的输出为：

```
[Car[B12, 18], Car[A12, 20], Car[Z34, 17], Car[Z34W, 18], Car[12NN, 19], Car[12N, 20]]
[Car[Z34, 17], Car[B12, 18], Car[Z34W, 18], Car[12NN, 19], Car[A12, 20], Car[12N, 20]]
Car[A12, 20]
Position of the searched car = 4
```

从输出结果看，排序与查找都是基于相同的语义来比较对象的。

总之，若想使用 Collections 中的算法对一组对象进行比较，那么应确保该对象的类实现了 Comparable 接口。如果对象的类型是自定义类，则还要注意保持 compareTo 方法、equals 方法和 hashCode 方法的一致性。

有的情况下需要按照字典顺序比较字符串，而有的情况下需要仅仅比较字符串的长度，即含有字符个数多的字符串大于含有字符个数少的字符串。这就需要针对特殊应用场景指定比较算法，而不是去修改 Comparable 的实现类。java.util.Comparator 接口就是用来建立对象比较算法与算法的使用者之间的接口。java.util.Comparator 接口中约定了抽象方法 compare 实现对象比较算法：int compare(T o1, T o2)，当第一个参数小于、等于或大于第二个参数时，该方法返回负整数、零或正整数。

源文件 7-9 演示了如何使用接口 Comparator 为 Collection.sort 指定对象比较算法，而不是使用默认自然排序。

源文件 7-9　ComparatorDemo.java

```
1 import java.util.Comparator;
2 import java.util.Collections;
3 import java.util.List;
4 import java.util.ArrayList;
5
6 class StringLengthComparator implements Comparator<String> {
```

```
 7      public int compare(String a, String b) {
 8          return a.length() - b.length();
 9      }
10 }
11
12 public class ComparatorDemo {
13     public static void main(String[] args) {
14         String[] s = {"abc", "ABC", "ab", "1234", "xyz"};
15         List<String> list = new ArrayList<>();
16         for (String item : s) {
17             list.add(item);
18         }
19         System.out.println(list);
20
21         Collections.sort(list, new StringLengthComparator());
22         System.out.println(list);
23     }
24 }
```

源文件 7-9 中 StringLengthComparator 类实现了 Comparator 接口，其中第 7 行的 compare 方法定义了如何使用字符串长度比较两个字符串大小。ComparatorDemo 类中的 main 方法首先在第 14 行到第 18 行用字符串数组创建了一个元素类型为 String 的线性表，第 21 行创建 StringLengthComparator 类型的对象，并将其作为参数调用 Collections.sort 方法对线性表排序。程序的输出是：

```
[abc, ABC, ab, 1234, xyz]
[ab, abc, ABC, xyz, 1234]
```

注意：第 6 行声明 StringLengthComparator 类实现 Comparator 接口时需为接口指定类型参数 String，否则 compare 方法形式参数就不能为 String 类型。

JDK 7 及以上版本的 Comparator 实现要满足如下三个条件，否则 Arrays.sort 和 Collections.sort 会抛出 IllegalArgumentException 异常：

(1) signum(compare(x,y))==-signum(compare(y,x));

(2) 如果((compare(x,y)>0)&&(compare(y,z)>0))那么 compare(x,z)>0；

(3) 如果 compare(x,y)==0 那么 signum(compare(x,z))==signum(compare(y,z))。

其中，如果参数为负数，则 signum 返回值为 −1；如果参数为零，则返回值为 0；如果参数为正，则返回值为 1。下例使用学生的年龄比较大小，但没有处理相等的情况，实际使用中可能会出现异常：

```
new Comparator<Student>() {
    @Override
    public int compare(Student o1, Student o2) {
        return o1.getAge() > o2.getAge() ? 1 : -1;
    }
};
```

对修改是封闭的；而对扩展是开放的。这称为程序设计的"开闭原则"，Comparator 接口的应用是符合开闭原则的，在设计模式中称为"策略模式"。

7.4 队列接口和实现类

队列是具有先进先出(FIFO)约束的线性表。java.util.Queue 接口定义了针对队列的一组操作。Java SE 中的队列接口 Queue<E>并没有继承 List<E>接口，而是直接继承了 Collection 接口。针对队列，一般需要入队、出队和查看队首元素等操作。java.util.Queue 接口定义这 3 个操作的两组抽象方法：一组抽象方法以抛出异常方式处理遇到的问题，另外一组以返回状态值的方式处理遇到的问题，如表 7-4 所示。

表 7-4 队列上的两类操作

操 作	抛出异常	返回状态值
入队	add(e)	offer(e)
出队	remove()	poll()
查看队首元素	element()	peek()

在固定容量的队列实现中，offer 方法可能会由于容量限制而返回 false；poll 方法可能由于队列为空而返回 null；add 方法在达到容量限制时抛出 IllegalStateException 异常；remove 方法在队列为空时抛出 NoSuchElementException 异常。ArrayBlockingQueue 类是 Queue 接口固定容量的实现类，这样的队列称为有界(bounded)队列。LinkedList 类实现了 Queue 接口，因此可以把 LinkedList 当成 Queue 来用。LinkedList 类并不限制元素的个数，这样的队列称为无界(unbounded)队列。

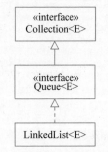

图 7-4 LinkedList 实现类

PriorityQueue 也是队列接口的一个实现类，队列中的元素默认按自然顺序(natural ordering)排序，Comparable 接口约定了对象间的自然顺序。因此当调用 peek 方法或者 poll 方法来取出队列中的元素时，并不是取出最先进入队列的元素，而是取出队列中优先级最高的元素。

图 7-4 展示了 Collection 接口，Queue 接口和 LinkedList 实现类之间的关系：Queue 接口继承了 Collection 接口；LinkedList 实现 Queue 接口。

假设有 3 个人依次来到游乐园售票窗口：张三、李四、王五。那么实际上他们就依次排在一个队列中。当开始售票后，张三出队，买票；李四出队，买票；最后是王五出队，买票。此时队列为空。源文件 7-10 模拟了排队买票的场景。

源文件 7-10　QueueDemo.java

```
1 import java.util.LinkedList;
2 import java.util.Queue;
3
4 public class QueueDemo {
5     public static void main(String[] args) {
6         String client;
7         Queue<String> queue = new LinkedList<>();
```

```
8        queue.add("张三");
9        queue.add("李四");
10       queue.add("王五");
11       System.out.println(queue);
12       client = queue.remove();
13       System.out.println(client + "买票。");
14       client = queue.remove();
15       System.out.println(client + "买票。");
16       client = queue.remove();
17       System.out.println(client + "买票。");
18       System.out.println(queue.isEmpty());
19     }
20 }
```

从第 7 行到第 11 行形成了排队买票的队列；从第 13 行到第 17 行让队列中的游客依次出队买票，直到队列为空。第 18 行判断队列是否为空。

程序的输出为：

```
[张三, 李四, 王五]
张三买票。
李四买票。
王五买票。
true
```

7.5 栈

栈是先进后出（LIFO）的线性表。Java SE 并没有为栈定义接口。由于双端队列 java.util.Deque 能够在线性表的两端进行插入和删除，所以使用 Deque 接口及其实现类能够实现栈的操作。Deque 继承了 Queue 接口，增加了在两端都添加和移除元素的功能。java.util.LinkedList 是 Deque 接口的一个实现类，另外还有一个专门高效的实现类 java.util.ArrayDeque。ArrayDeque 比 java.util.LinkedList 和 java.util.Stack 快，而且没有容量限制。建议使用 ArrayDeque 实现栈而不是 Stack。接口 Deque 为栈操作定义了 push、pop 和 peek 方法，如表 7-5 所示；为队列操作定义的 add、remove 和 element 方法如表 7-6 所示。

表 7-5 接口 Deque 中的栈操作

栈方法	等价的双端队列方法	功能
push	addFirst	入栈，将对象压入栈顶
pop	removeFirst	弹栈，将栈顶对象移出并返回
peek	getFirst	查看栈顶对象，但并不移出

表 7-6 接口 Deque 中的队列操作

队列方法	等价的双端队列方法	功能
add	addLast	入队，将对象排在队尾
remove	removeFirst	出队，将队首对象移出
element	getFirst	查看队首元素，但并不移出

图 7-5 展示了 Queue、Deque 和 ArrayDeque 之间的关系：Deque 接口是 Queue 接口的子接口，ArrayDeque 实现了 Deque。

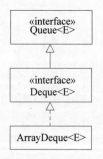

图 7-5 栈的接口及实现类

源文件 7-11 演示了如何把双端队列作为栈来使用。从第 7 行到第 12 行创建了元素类型为 Integer 的双端队列作为栈使用，并使用 push 方法把整数 5、67、789 入栈，第 13 行输出此时栈中的内容。第 16 行使用 pop 方法出栈、第 18 行再出栈，第 20 行把整数 123 入栈，第 21 行输出栈中的内容。第 22 行查看栈顶元素。

源文件 7-11　StackDemo.java

```java
1  import java.util.Deque;
2  import java.util.ArrayDeque;
3
4  public class StackDemo {
5      public static void main(String[] args) {
6          //用 ArrayDeque 作为栈
7          Deque<Integer> stack = new ArrayDeque<>();
8
9          //入栈
10         stack.push(5);
11         stack.push(67);
12         stack.push(789);
13         System.out.println(stack);
14
15         //出栈入栈
16         stack.pop();
17         System.out.println("Stack after pop: " + stack);
18         stack.pop();
19         System.out.println("Stack after pop: " + stack);
20         stack.push(123);
21         System.out.println("Stack after push: " + stack);
22         System.out.println("Stack top: " + stack.peek());
23     }
24 }
```

程序的输出为：

```
[789, 67, 5]
Stack after pop: [67, 5]
Stack after pop: [5]
Stack after push: [123, 5]
Stack top: 123
```

7.6 Map 接口和实现类

　　Set 接口和 List 接口是 Collection 接口两个常用的子接口。在 JCF 框架中，还有一个与 Collection 接口同样重要的接口：Map。一个 Map 类型的集体就是"键-值"对（key-value pairs）的集合，在这个集合中基于"键"存储查找"值"。无论"键"还是"值"，都必须是对象。"键"就意味着不允许对象重复。Map 接口定义了"键-值"对集合上的操作。

　　图 7-6 展示 Map 集体的概念。图中的钥匙表示"键"对象，椭圆表示"值"对象，从钥匙到椭圆的箭头表示映射关系。键是能够唯一标识"值"对象的属性。比如，可以使用字符串对象"车牌号码"来唯一标识一个轿车对象。

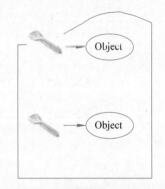

图 7-6　Map 集体是键-值对的集合

　　Map 集体上最基本的操作就是存入和取出，这两个操作在 Map 接口中定义为抽象方法 put 和 get。对于 put 方法而言，应当接收一个"键-值"对作为其参数。所以 put 方法中有两个参数。由于这两个参数的类型也需要客户程序指定，所以这里采用了类型参数 K 和 V。类型参数 K 和 V 声明在接口名字的后面，括在一对尖括号"< >"中，使用逗号隔开。

　　从集体中存入（put）对象，从集体中取出（get）指定对象，删除（remove）指定对象，判断集体中是否含有指定键（containsKey），判断集体中是否含有指定的值（containsValue），集体中元素的个数（size），集体是否为空（isEmpty），返回集体中键的集合（keySet），返回集体中值的集合（values）等操作的抽象方法详细定义如下：

```
public interface Map<K,V> {

    //基本操作
    //添加"键-值"对
    V put(K key, V value);
    //由参数指定的键返回相应的值
    V get(Object key);
    //删除参数指定键对应的"键-值"对
    V remove(Object key);

    //如果含有参数指定的键返回 true
    boolean containsKey(Object key);
    //如果参数指定的值至少出现一次,则返回 true
    boolean containsValue(Object value);
    //返回"键-值"对的实际数目
    int size();
    //如果没有任何"键-值"对,则返回 true
    boolean isEmpty();
    //删除所有"键-值"对
```

```
        void clear();

        //集体视图
        //把所有"键"放在集合中返回
        public Set<K> keySet();
        //把所有"值"放在集体中返回
        public Collection<V> values();

        //Map中元素键-值对的接口类型
        public interface Entry {
            //返回"键"
            K getKey();
            //返回"值"
            V getValue();
            //设置"值"
            V setValue(V value);
        }
    }
```

注意：在上面定义中，还定义了作为元素的"键-值"对接口：Entry。Entry接口约定了方法访问Map集体中元素对象的方法：返回元素中的"键"（getKey），返回元素中的"值"（getValue），设置元素中的"值"（setValue）。

JCF框架中，Map接口有两个实现类：HashMap和TreeMap。如果不关心元素次序，使用HashMap；否则，使用TreeMap。HashMap是Map接口的基于哈希表的实现。HashMap集体允许存放空键和空值。试图作为HashMap集体中"键"的对象的类型中必须实现hashCode方法和equals方法。这是因为put方法已经被约定去调用对象上的hashCode方法计算其在哈希表中的位置，约定使用equals方法判断两个键是否重复。

图7-7展示了HashMap的存储结构。其中，每个小的矩形框表示一个"键-值"对元素。大的矩形框表示哈希表。如果两个键的HashCode返回值相等，则会产生地址冲突。链地址法将冲突的"键-值"对存储在一个单链表中。

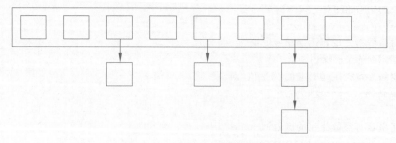

图7-7　HashMap的存储结构

当使用put方法向Map集体中存放一个"键-值"对时，put方法首先调用"键"对象上的hashCode方法来计算其在哈希表中的存放位置。hashCode方法的返回值并不能直接作为存放位置。假设哈希表的容量是$N(2^n)$，那么整个哈希表的元素位置索引范围是[0, N)。put方法应用一个映射函数把hashCode方法返回的对象哈希码映射到[0, N)中的某个值。

有两个重要的参数会影响哈希表的性能：初始容量（initial capacity）和装载因子（load

factor)。装载因子是哈希表饱和程度的度量：length/capacity，其中 length 是哈希表中已有的元素个数，capacity 是哈希表的容量。默认的装载因子是 0.75，即当哈希表中有 75% 的单元被使用后，put 方法会再增加哈希表的容量。源文件 7-12 演示了 Map 接口及实现类 HashMap 的使用。

源文件 7-12　HashMapDemo.java

```
1   import java.util.Map;
2   import java.util.HashMap;
3   import java.util.Set;
4
5   public class HashMapDemo {
6       public static void main(String[] args) {
7           Car a = new Car("A12", 18);
8           Car b = new Car("B12", 20);
9           Car c = new Car("C12", 17);
10          Car d = new Car("D56", 18);
11          Car e = new Car("E56", 19);
12          Car f = new Car("F56", 20);
13          Car ff = new Car("F56", 21);
14
15          Map<String, Car> map = new HashMap<>();
16
17          map.put("A12", a);
18          map.put("B12", b);
19          map.put("C12", c);
20          map.put("D56", d);
21          map.put("E56", e);
22          map.put("F56", f);
23          map.put("F56", ff);
24
25          //输出对象的字符串表示
26          System.out.println(map);
27
28          //remove 方法按照键删除键-值对,返回键所标识的对象(值)
29          //如果参数并不是集体中的键,则返回 null
30          Car s3 = map.remove("C12");
31          System.out.println(s3 + " has been removed.");
32
33          //根据键检索值
34          Car s4 = map.get("D56");
35
36          //根据键判断是否包含
37          System.out.println(map.containsKey("C12")
38              ? "C12 is there." : "C12 is not there.");
39          System.out.println(map.containsValue(d)
40              ? "D56 is there." : "D56 is not there.");
41
42          System.out.println("The amount of key-value mappings is "
43              + map.size());
44
```

```java
45          //根据"键"遍历
46          Set<String> s = map.keySet();
47          for (String key : s) {
48              Car value = map.get(key);
49              System.out.print("(" + key + ", " + value + ") ");
50          }
51          System.out.println();
52
53          //根据"值"遍历
54          for (Car value : map.values()) {
55              System.out.print(value + " ");
56          }
57          System.out.println();
58
59          //根据"键-值"对遍历
60          for (Map.Entry<String, Car> entry : map.entrySet()) {
61              String key = entry.getKey();
62              Car value = entry.getValue();
63              System.out.print("(" + key + ", " + value + ")   ");
64          }
65          System.out.println();
66
67          //清空
68          map.clear();
69          System.out.println(map.isEmpty()
70              ? "Map is empty."  : "Map is not empty.");
71      }
72  }
73
74  class Car {
75      private String licenceNumber;
76      private Integer weight;
77
78      Car(String name, int weight) {
79          this.licenceNumber = name;
80          this.weight = weight;
81      }
82
83      @Override
84      public String toString() {
85          return "Car [" + licenceNumber + ", " + weight + "]";
86      }
87
88      /*
89       * (non-Javadoc)
90       *
91       * @see java.lang.Object#hashCode()
92       */
93      @Override
94      public int hashCode() {
95          final int prime = 31;
96          int result = 1;
```

```
 97         result = prime * result + ((licenceNumber == null)
 98             ? 0 : licenceNumber.hashCode());
 99         result = prime * result + ((weight == null)
100             ? 0 : weight.hashCode());
101         return result;
102     }
103
104     /*
105      * (non-Javadoc)
106      *
107      * @see java.lang.Object#equals(java.lang.Object)
108      */
109     @Override
110     public boolean equals(Object obj) {
111         if (this == obj) {
112             return true;
113         }
114         if (obj == null) {
115             return false;
116         }
117         if (getClass() != obj.getClass()) {
118             return false;
119         }
120         Car other = (Car) obj;
121         if (licenceNumber == null) {
122             if (other.licenceNumber != null) {
123                 return false;
124             }
125         } else if (! licenceNumber.equals(other.licenceNumber)) {
126             return false;
127         }
128         if (weight == null) {
129             if (other.weight != null) {
130                 return false;
131             }
132         } else if (! weight.equals(other.weight)) {
133             return false;
134         }
135         return true;
136     }
137 }
```

在第 74 行开始的类 Car 的定义中，声明了具有车牌号码和车身质量两个成员变量、一个有参数的构造方法，覆盖了 toString 方法。然后声明覆盖方法 hashCode 和 equals。

在第 5～11 行类 HashMapDemo 的 main 方法中先创建了 7 个 Car 类型的对象，分别使用引用变量 a、b、c、d、e、f 和 ff 引用。其中，对象 ff 和 f 的键相同。第 17～23 行使用 put 方法向 Map 对象中放入这 7 个对象，第 24 行输出是：

```
{A12=Car [A12, 18], B12=Car [B12, 20], D56=Car [D56, 18],
C12=Car [C12, 17], F56=Car [F56, 21], E56=Car [E56, 19]}
```

可以看到，Map 对象中只有 6 个元素。语句 map.put("F56", ff) 并没有把键-值对（字符串对象 "F56"，ff 引用的 Car 对象）存放到 Map 对象中。注意，"键"的类型是 String，"值"

的类型是 Car。

第 30 行调用实例方法 remove 把键为"C12"的元素删除；第 34 行调用实例方法 get 查询键为"D56"的"值"。第 37 行调用实例方法 containsKey 判断键为"C12"的元素是否存在，结果显示不存在：

```
C12 is not there.
```

第 39 行调用实例方法 containsValue 判断是否存在 d 所引用的对象"(D56,18)"，程序输出：

```
D56 is there.
```

第 42 行输出 Map 集体中的元素总数：

```
The amount of key-value mappings is 5
```

接下来使用三种方法遍历 Map 集体。第 46 行到第 50 行根据"键"遍历 Map 集体中的所有元素：首先使用 keySet 方法获取所有的键，放到一个集合 s 中；然后使用 for-each 循环遍历集合 s：对于集合 s 中的每一个键 key，让集体对象使用 get 方法返回 key 映射的值对象 value。然后访问值对象 value。遍历的结果是：

```
(A12, Car[A12, 18])  (B12, Car[B12, 20])  (D56, Car[D56, 18])  (F56, Car[F56, 21])  (E56, Car[E56, 19])
```

第 54 行到第 56 行根据"值"遍历，结果为：

```
Car[A12, 18] Car[B12, 20] Car[D56, 18] Car[F56, 21] Car[E56, 19]
```

第 60 行到第 64 行根据"键-值对"遍历。首先使用实例方法 entrySet() 把所有元素放入一个集合 Set 中，一个元素就是一个键-值对，其类型为 Entry。Entry 是 Map 接口中声明的接口。然后使用 for-each 循环遍历这个集合：调用 Entry 对象的实例方法 getKey() 查询元素的键；调用 Entry 对象的实例方法 getValue() 查询元素的值。

第 68 行调用实例方法 clear 清除所有元素。第 69 行再调用实例方法 isEmpty 判断是否为空。程序输出：

```
Map is empty.
```

注意，本例中完整的程序输出为：

```
{A12=Car[A12, 18], B12=Car[B12, 20], D56=Car[D56, 18], C12=Car[C12, 17], F56=Car[F56, 21], E56=Car[E56, 19]}
Car[C12, 17] has been removed.
C12 is not there.
D56 is there.
The amount of key-value mappings is 5
(A12, Car[A12, 18])  (B12, Car[B12, 20])  (D56, Car[D56, 18])  (F56, Car[F56, 21])  (E56, Car[E56, 19])
Car[A12, 18] Car[B12, 20] Car[D56, 18] Car[F56, 21] Car[E56, 19]
(A12, Car[A12, 18])  (B12, Car[B12, 20])  (D56, Car[D56, 18])  (F56, Car[F56, 21])  (E56, Car[E56, 19])
Map is empty.
```

注意：HashMap 使用 String 类的 hashCode 方法和 equlas 方法计算键-值对的哈希地

址和判断键是否重复。如果使用自定义类型作为键,则该类型必须实现 hashCode 方法和 equlas 方法。

7.7 流

在使用集体处理大数据的场景中,往往需要根据某些条件过滤数据。流(stream)是从来源(source)到归宿(sink)的元素序列(sequence of elements)。流的最大特征是没有存储。早晨上班时汽车从小区车库开出驶入公路形成车流;到达后停在公司停车场。"小区车库"和"公司车库"相当于集体对象等数据存储;车流相当于元素流。从 Java 8 开始新增了包 java.util.stream 来解决对大数据的集体操作,例如过滤(filter)、map/reduce、聚合等,称为流 API。流 API 既可以串行执行,也可以并行执行。

在 Java 8 之前,如果想对一个整数线性表中不小于 90 的元素求和,那么会设计如下程序:

```
private static int sumList(List<Integer> list) {
    Iterator<Integer> it = list.iterator();
    int sum = 0;
    while (it.hasNext()) {
        int num = it.next();
        if (num >= 90) {
            sum += num;
        }
    }
    return sum;
}
```

为了得到满足条件元素的总和,这个程序需要程序员了解过程细节:获取迭代器、迭代处理线性表等。这种做法称为"外部迭代"。流 API 的做法是"内部迭代",使得程序更加简洁可读:

```
private static int sumList(List<Integer> list) {
    return list.stream().filter(p -> p >= 90).mapToInt(p -> p).sum();
}
```

其中,方法 filter 的参数 p->p>=90 读作:以每个元素 p 为参数计算关系表达式 p>=90。这样就实现了按照条件 p>=90 对流中的数据过滤。过滤后的流再作为 mapToInt 方法的输入,把每个元素变换为 Integer 对象,最后对这些整数对象求和。源文件 7-13 展示了使用流 API 计算一组成绩中 90 及 90 以上成绩的平均数。第 15 行 scores.stream() 创建了一个元素为 Double 对象的流。第 16 行使用 Lambda 表达式设置过滤条件,由于第 15 行的输出是第 16 行的输入,所以第 16 行中形式参数 p 的实际参数就是流中的每一个 Double 对象。filter()称为中间操作(intermediate operations),该方法按照给定条件过滤每个元素。

第 17 行 mapToDouble(Double::doubleValue) 把每个元素都视为 double 类型进行处理。虽然对于 Double 对象来说这一步不必要,但在处理 Integer 或 String 等需要转换成 double 时会用到。在 Java 8 及以后的版本中,Double::doubleValue 是一个方法引用

(Method Reference),它引用了 java.lang.Double 类中的静态方法 doubleValue(double),:: 称为方法引用操作符。一般使用中间操作.map()在每个元素上应用给定的函数。

第 18 行中 average() 计算平均值并返回一个 Optional 对象,这是因为如果流为空,就没有平均值可言。第 19 行中的 orElse(0.0) 表示如果产生平均值则返回该值,否则返回默认值 0.0。

源文件 7-13 StreamAverageCalculator.java

```java
1 import java.util.List;
2 import java.util.ArrayList;
3
4 public class StreamAverageCalculator {
5     public static void main(String[] args) {
6         //创建一个包含成绩的线性表
7         List<Double> scores = new ArrayList<>();
8         scores.add(87.5);
9         scores.add(90.0);
10        scores.add(78.5);
11        scores.add(97.0);
12        scores.add(95.0);
13
14        //使用 Stream API 计算 90 及 90 以上平均值
15        double averageScore = scores.stream()
16                              .filter(p -> p >= 90)
17                              .mapToDouble(Double::doubleValue)
18                              .average()
19                              .orElse(0.0); //如果列表为空,则返回默认值 0.0
20
21        System.out.println("平均成绩是: " + averageScore);
22    }
23 }
```

程序的输出为:

平均成绩是: 94.0

Lambda 表达式是单抽象方法接口的实现类的实例。最简单的 Lambda 表达式形如:

(<参数>) -> <表达式>

意思是根据<参数>来计算表达式<表达式>。如果需要两个参数 parameter1 和 parameter2,那么 Lambda 表达式形如:

(parameter1, parameter2) -> <表达式>

一般地,完成复杂计算的多个表达式使用一对花括号{}括起来,Lambda 表达式形如:

(parameter1, parameter2, ……) -> {<代码块>}

当<代码块>中只有一个语句时,可以省略花括号{}。

一个 Lambda 表达式是一个单抽象方法接口匿名实现类的实例。单抽象方法接口就是仅有一个抽象方法的接口,如:

```java
interface IStudent {
    String getName();
}
```

假设其实现类为：

```
class StudentImpl implements IStudent {
    String name;
    @Override
    public String getName() {
        return name;
    }
}
```

那么客户程序就可以通过接口声明引用变量，通过实现类创建对象：

```
IStudent a = new StudentImpl();
a.getName();
```

而使用 Lambda 表达式，则无须声明 StudentImpl 实现类：

```
IStudent a = () -> {
    return name;
};
a.getName();
```

这样使得程序更加简洁并易于扩展。

通过创建一个 Stream 对象，并在其上连续应用一系列中间操作（Intermediate Operations）和终结操作（Terminal Operations），形成一个数据处理管道，称为"管道式处理"，也就是 Stream API 的链式调用。中间操作包括筛选 filter()、排序 sort()、映射 map() 等，中间操作生成一个新的 Stream 对象以进行下一步的处理。终结操作包括遍历 forEach()、查找 findFirst()、聚合 count()、归约 reduce()、收集 collect() 等。除了 count()，聚合操作还有 max()、min()、sum()、average() 等。终结操作会触发实际的计算，并产生最终结果或者副作用，完成终结操作后流就终止了。如源文件 7-13 第 16 行到第 19 行所示。

源文件 7-14 演示了流的创建、终结操作、串行流和并行流等。第 9 行通过类方法 Stream.of 创建流对象。第 10 行、第 11 行、第 13 行、第 15 行分别在流上应用终结操作 count、forEach、anyMatch、reduce。注意这些终结操作应用于不同的流对象，尽管元素序列相同。

第 19 行准备了一个线性表存放 100 个整数，第 24 行从该线性表创建了串行流对象。第 29 行从该线性表创建了并行流对象。从输出结果可以看到，在串行流上过滤的结果与并行流上过滤的结果不同：前者保留了顺序而后者没有保留顺序。

源文件 7-14　StreamDemo.java

```
1 import java.util.ArrayList;
2 import java.util.List;
3 import java.util.stream.Stream;
4 import java.util.Optional;
5
6 public class StreamDemo {
7
8     public static void main(String[] args) {
9         Stream<Integer> numbers = Stream.of(1, 2, 3, 4, 5);
10        System.out.println("Number of elements = " + numbers.count());
11        Stream.of(1, 2, 3, 4, 5).forEach(i -> System.out.print(i + ","));
```

```
12        System.out.println("Stream contains 4? "
13            + Stream.of(1, 2, 3, 4, 5).anyMatch(i -> i == 4));
14        Optional<Integer> optionalResult
15            = Stream.of(1, 2, 3, 4, 5).reduce((i, j) -> { return i * j; });
16        if (optionalResult.isPresent()) System.out.println(
17            "accumulative multiplication = " + optionalResult.get());
18
19        List<Integer> list = new ArrayList<>();
20        for (int i = 0; i < 100; i++) {
21            list.add(i);
22        }
23        //从串行流过滤不小于 90 的元素
24        Stream<Integer> sequentialStream = list.stream();
25        Stream<Integer> t = sequentialStream.filter(p -> p >= 90);
26        t.forEach(p -> System.out.print(p + " "));
27        System.out.println();
28        //从并行流过滤不小于 90 的元素
29        Stream<Integer> parallelStream = list.parallelStream();
30        Stream<Integer> s = parallelStream.filter(p -> p >= 90);
31        s.forEach(p -> System.out.print(p + " "));
32        System.out.println();
33    }
34 }
```

程序的输出为：

```
Number of elements = 5
1,2,3,4,5,Stream contains 4? true
accumulative multiplication = 120
90 91 92 93 94 95 96 97 98 99
90 91 92 96 97 98 99 93 94 95
```

当希望以线性表返回流中所有元素时，可使用操作：

```
stream.collect(Collectors.toList())
```

其中：

stream 是一个流对象，可能来自一个数组、集合或其他数据源；

collect 是流 API 中的"收集"操作；

Collectors.toList 是 Collectors 类的一个静态方法，返回一个 Collector 接口实现类的对象，该对象定义了如何把流转换为线性表。

例如从三个字符串 a，ab，cab 中查找以 a 打头的字符串，就可以先把存放这三个字符串的线性表转换为流，然后通过流上的过滤和收集操作以线性表返回结果。如源文件 7-15 所示。

源文件 7-15　CollectorsDemo.java

```
1 import java.util.List;
2 import java.util.ArrayList;
3 import java.util.stream.Collectors;
4
5 public class CollectorsDemo {
6     public static void main(String[] args) {
```

```
 7        List<String> list = new ArrayList<>();
 8        list.add("a");
 9        list.add("ab");
10        list.add("cab");
11        List<String> result = list.stream()
12                   .filter(s -> s.startsWith("a"))
13                   .collect(Collectors.toList());
14        System.out.println(result);
15    }
16 }
```

程序的输出为：

[a, ab]

注意：Stream 自己不存储元素对象；Stream 不改变源，每次操作都会返回一个持有结果的新 Stream。中间操作可以是零个或多个；终结操作至少一个。

终结操作通常返回一个 Optional 对象。Optional 的意思是"可能有，也可能没有"。Optional 提高了代码的健壮性和可读性。例如如果 Optional 对象可能为 null，则可使用 Optional.ofNullable()来创建 Optional 对象：

```
Optional<String> optional = Optional.ofNullable(null);
```

如果值存在，那么 isPresent()方法会返回 true；此时可调用 get()方法获取返回值：

```
if(optional.isPresent()) {
   System.out.println(optional.get());
} else {
   System.out.println("没有");
}
```

也可以使用 orElse 方法。例如如果 Optional 集体对象中有值 orElse()则返回该值，否则返回字符串"没有"：

```
String result = optional.orElse("没有");
```

例如已知整数序列：7,5,2,9,3,8,6,1，计算大于 6 的整数有哪些。程序如源文件 7-16 所示。该程序首先从一个整数数组创建一个线性表，然后使用线性表的实例方法 stream 创建流对象。调用流对象上的 filter 方法过滤大于 6 的整数，逐个输出。然后在过滤流上调用 findFirst 方法查找第一个大于 6 的整数，结果放在 Optional 对象中。第 23 行调用 Optional 对象的 get 方法获取查找的结果，即第一个大于 6 的数。第 19 行从集体对象创建了并行流，然后过滤，最后返回任意一个满足条件的整数。

源文件 7-16　TestOptional.java

```
 1 import java.util.Optional;
 2 import java.util.stream.Stream;
 3 import java.util.List;
 4 import java.util.ArrayList;
 5
 6 public class TestOptional {
 7    public static void main(String[] args) {
 8        int[] a = {7,5,2,9,3,8,6,1};
 9        List<Integer> list = new ArrayList<>();
```

```
10        for (int e : a) {
11            list.add(e);
12        }
13        //遍历输出符合条件的元素
14        list.stream().filter(x -> x > 6).forEach(System.out::println);
15        //查找第一个符合条件的元素
16        Optional<Integer> findFirst = list.stream()
17            .filter(x -> x > 6).findFirst();
18        //查找任意一个符合条件的元素
19        Optional<Integer> findAny = list.parallelStream()
20            .filter(x -> x > 6).findAny();
21        //判断是否包含符合特定条件的元素
22        boolean anyMatch = list.stream().anyMatch(x -> x > 6);
23        System.out.println("第一个值: " + findFirst.get());
24        System.out.println("任意一个值: " + findAny.get());
25        System.out.println("是否存在: " + anyMatch);
26    }
27 }
```

程序的输出为：

```
7
9
8
第一个值: 7
任意一个值: 8
是否存在: true
```

源文件 7-17 展示了如何按照自然顺序查找一组整数中的最大值，以及如何按照字符串查找一组整数中的最大整数。程序首先把一组整数由数组转换为线性表，然后从线性表创建流对象。在流对象上应用 max() 方法查找最大元素。第 15 行以 Integer 类中的 compareTo 方法作为 max() 方法调用实际参数，意思是按照 Comparable 接口的 compareTo 方法进行比较（即按自然顺序），查找的结果存放在 Optional 对象中。第 24 行则从这个 Optional 对象中获取结果。第 16 行则通过给 max() 方法传入一个 Comparator 对象来改变排序策略：按照字符串排序。

源文件 7-17　TestMax.java

```
1 import java.util.Optional;
2 import java.util.stream.Stream;
3 import java.util.List;
4 import java.util.ArrayList;
5 import java.util.Comparator;
6
7 public class TestMax {
8     public static void main(String[] args) {
9         int[] a = {7, 25, 2, 9, 3, 8, 6, 54};
10        List<Integer> list = new ArrayList<>();
11        for (int e : a) {
12            list.add(e);
13        }
14        //自然排序
```

```
15      Optional<Integer> max = list.stream().max(Integer::compareTo);
16      //自定义排序
17      Optional<Integer> max2 = list.stream()
18          .max(new Comparator<Integer>() {
19              @Override
20              public int compare(Integer a, Integer b) {
21                  return a.toString().compareTo(b.toString());
22              }
23          });
24      System.out.println("按整数排序的最大值: " + max.get());
25      System.out.println("按字符串排序的最大值: " + max2.get());
26   }
27 }
```

程序的输出为：

按整数排序的最大值: 54
按字符串排序的最大值: 9

map()就是把某函数应用到流的每个元素上，映射成新的元素，形成新流。这是一个一对一的转换操作，不会改变流的长度，只是改变元素的内容或类型。用于将流中的元素组合起来，生成一个单一的结果。reduce()则用来实现从多个元素到一个单一值的聚合操作，比如求和、求最大值、求最小值等。

源文件 7-18 展示了如何把流中的每个元素 x 映射到 x+2。

源文件 7-18　TestMap.java

```
1 import java.util.stream.Stream;
2 import java.util.stream.Collectors;
3 import java.util.List;
4 import java.util.ArrayList;
5
6 public class TestMap {
7    public static void main(String[] args) {
8        int[] a = {7,25,2,9,3,8,6,54};
9        List<Integer> list = new ArrayList<>();
10       for (int e : a) {
11           list.add(e);
12       }
13       List<Integer> listNew = list.stream().map(x -> x + 2)
14           .collect(Collectors.toList());
15       System.out.println("每个元素+2: " + listNew);
16   }
17 }
```

程序的输出为：

每个元素+2: [9, 27, 4, 11, 5, 10, 8, 56]

注意：因为流没有数据存储，那么在流中的数据完成处理后，需要将流中的数据重新归集到新的集合对象里。Collectors.toList、Collectors.toSet 和 Collectors.toMap 是比较常用的归集方法。第 14 行就是使用 Collectors.toList 把流对象归集到一个线性表对象。

源文件 7-19 展示了如何对一组整数求和和求最大值。求和、求最大值都是终结操作。

第 15 行通过 Lambda 表达式进行归约操作，这个 Lambda 表达式定义了二元加法运算；第 18 行则是通过引用 Integer.sum 方法实现归约。从第 27 行的输出看，二者结果相同。第 22 行通过 Lambda 表达式进行归约操作实现比较；第 25 行则是通过方法调用定义二元比较运算，将连续的两个元素进行比较，取两者中的较大值。这意味着整个 Stream 会被遍历，每一步都将当前的最大值与下一个元素比较，选取更大的那个作为新的最大值。

源文件 7-19　TestReduce.java

```java
1   import java.util.List;
2   import java.util.ArrayList;
3   import java.util.stream.Stream;
4   import java.util.stream.Collectors;
5   import java.util.Optional;
6
7   public class TestReduce {
8       public static void main(String[] args) {
9           List<Integer> list = new ArrayList<>();
10          list.add(1); list.add(2); list.add(3);
11          list.add(4); list.add(5);
12
13          //通过 Lambda 表达式归约求和
14          Optional<Integer> sumByExp = list.stream()
15              .reduce((x, y) -> x + y);
16          //通过方法调用归约求和
17          Optional<Integer> sumByMethod = list.stream()
18              .reduce(Integer::sum);
19
20          //通过 Lambda 表达式归约求最大值
21          Optional<Integer> maxByExp = list.stream()
22              .reduce((x, y) -> x > y ? x : y);
23          //通过方法调用求最大值
24          Optional<Integer> maxByMethod = list.stream()
25              .reduce((x, y) -> Integer.max(x, y));
26
27          System.out.println("和: " + sumByExp.get() + ", "
28              + sumByMethod.get());
29          System.out.println("最大值: " + maxByExp.get() + ", "
30              + maxByMethod.get());
31      }
32  }
```

程序的输出为：

和: 15, 15
最大值: 5, 5

第 7 章　章节测验

第 8 章 泛型

泛型(Generics)就是把类型作为参数,也称为参数化类型。由于泛型使得能够把类型作为参数传递给类、接口和方法,就使得程序对集体对象中元素的操作更加安全。

8.1 概述

在 Java SE 没有泛型之前,如果期望不同类型的对象放在同一个集体对象中,那么该集体对象中元素的类型只能是 Object。源文件 8-1 在第 6 行使用类 ArrayList 创建线性表。在 Java 世界中,所有的类都是 Object 的子类,所有类的对象都能够上转型到 Object 类型。因此,可以向这个线性表中添加任意类型的对象:String 类型的对象,或者 Integer 类型的对象,或者自定义类型的对象,等等。

第 7 行把整数 23 添加到线性表中,编译器会把整数字面量 23 装箱为 Integer 对象;第 8 行使用字符串对象 abc 添加到线性表中。第 9 行和第 10 行按照位置索引查询线性表中的元素并试图按整数类型输出元素的值。

源文件 8-1　NoGenericsDemo.java

```
1 import java.util.List;
2 import java.util.ArrayList;
3
4 public class NoGenericsDemo {
5     public static void main(String[] args) {
6         List list = new ArrayList();
7         list.add(23);
8         list.add("abc");
9         System.out.println(((Integer) list.get(0)).intValue());
10        System.out.println(((Integer) list.get(1)).intValue());
11    }
12 }
```

第 9 行中的 get(0)实例方法返回 Integer 类型的元素并输出其值：23。

第 10 行的 get(1)实例方法调用就出问题了。位置索引 1 上的对象是字符串对象"abc"，实例方法 get(1)返回该字符串对象，但当将这个字符串对象转型为 Integer 对象时就会抛出异常：

```
Exception in thread "main" java.lang.ClassCastException: class java.lang.
String cannot be cast to class java.lang.Integer (java.lang.String and java.
lang.Integer are in module java.base of loader 'bootstrap')
    at NoGenericsDemo.main(NoGenericsDemo.java:10)
```

这个异常是运行时刻异常 ClassCastException：不能把 java.lang.String 类型转型为 java.lang.Integer 类型。

无元素类型约束的集体对象操作使得直到程序在运行时刻才能发现问题。如果使用泛型，则可以把问题的发现提前到编译时刻，以语法错误的形式解决。源文件 8-2 演示使用泛型的版本。

源文件 8-2　GenericsDemo.java

```
1 import java.util.List;
2 import java.util.ArrayList;
3
4 public class GenericsDemo {
5     public static void main(String[] args) {
6         List<Integer> list = new ArrayList<Integer>();
7         list.add(23);
8         list.add("abc");        //参数类型不匹配
9         System.out.println(list.get(0).intValue());
10        System.out.println(list.get(1).intValue());
11    }
12 }
```

源文件 8-2 第 6 行使用 new ArrayList<Integer>创建线性表，并且指定元素 Integer 类型。这样的语句就告诉编译器，线性表中仅运行存放 Integer 类型的对象。所以，第 7 行调用实例方法 add 添加整数 23 没有问题。但是，第 8 行当试图添加字符串对象 abc 时，编译器就会提示：

```
GenericsDemo.java:8: error: incompatible types: String cannot be converted
  to Integer
```

意思是：线性表已经约定元素为 Integer 类型，不能使用形式参数为 Integer 类型的 add 方法添加 String 类型的实际参数。

这样，通过泛型，编译器就可以通过参数检查保证类型兼容，从而保证了类型安全。

如果希望线性表中存放任意数值对象，而不仅仅存放整数对象。也就是说，也允许 Float 类型、Double 类型的对象作为元素，可使用源文件 8-3 版本。

源文件 8-3　GenericsNumberDemo.java

```
1 import java.util.List;
2 import java.util.ArrayList;
3
4 public class GenericsNumberDemo {
5     public static void main(String[] args) {
6         List<Number> list = new ArrayList<Number>();
```

```
7        list.add(23);
8        list.add(3.14);
9        System.out.println(list.get(0).intValue());
10       System.out.println(list.get(1).intValue());
11   }
12 }
```

源文件 8-3 第 6 行使用类型参数 Number 创建了线性表：new ArrayList<Number>，从而声明了线性表中元素的类型必须是 Number。由于 Integer 类和 Double 类都是 Number 的子类，即这些类的对象都能自动转型到 Number 类型，所以这些类的对象允许放到线性表中，称为类型兼容的对象。

8.2 泛型类

在 JCF 框架中，所有的接口和类都是以泛型定义的。也可以定义自己的泛型类(Types of Generics)，形如：

```
class 类名 <类型参数 1, 类型参数 2,……> {
    //类体
}
```

声明了类型参数的类称为泛型类。类型参数声明在类名后面一对尖括号中。类型参数可以有多个，多个类型参数使用逗号分隔。泛型类型不能是基本数据类型。在 Java 中自定义的泛型类是为了创建可以处理多种数据类型的类，而不必为每种数据类型编写单独的类。泛型允许在类声明时指定一个或多个类型参数，然后在类中使用这些类型参数。源文件 8-4 是一个简单的泛型类定义示例。

源文件 8-4　MyGenericDemo.java

```
1 class MyGenericClass<T> {
2
3      //泛型类型 T 的实例变量
4      private T value;
5
6      //构造函数接收泛型类型 T 的参数
7      MyGenericClass(T value) {
8          this.value = value;
9      }
10
11     //获取泛型类型 T 的实例变量
12     public T getValue() {
13         return value;
14     }
15
16     //设置泛型类型 T 的实例变量
17     public void setValue(T value) {
18         this.value = value;
19     }
20 }
```

```
21
22  public class MyGenericDemo {
23      public static void main(String[] args) {
24          MyGenericClass<String> a = new MyGenericClass<>("Hello");
25          System.out.println(a.getValue());
26
27          MyGenericClass<Integer> b = new MyGenericClass<>(23);
28          System.out.println(b.getValue());
29      }
30  }
```

程序的输出为：

```
Hello
23
```

源文件 8-4 首先定义了泛型类 MyGenericClass，这个类有一个类型参数 T。类中有一个实例变量 value，其类型为类型参数 T。使用类型参数 T 还声明了有参数构造方法以及 getter 和 setter 方法。

第 24 行使用类型实际参数 String 创建了实例变量类型为 String 的类，并使用该类实例化了成员变量 value 值为字符串 Hello 的对象。第 25 行调用该对象上的实例方法 getValue 获取实例变量 value 的值。

第 27 行使用类型实际参数 Integer 创建了实例变量类型为 Integer 的类，并使用该类实例化了成员变量 value 值为整数 23 的对象。第 28 行调用该对象上的实例方法 getValue 获取实例变量 value 的值。

8.3 泛型接口

有泛型类，也有泛型接口。含类型参数的接口称作泛型接口，形如：

interface 名称<类型参数 1, 类型参数 2, ……>

例如对栈的操作有入栈、弹栈、查看栈顶元素、判断栈空、判断栈满、把栈清空等；栈中的元素可以是各种类型的对象。这就需要定义一个泛型接口 IStack<E>，如源文件 8-5 所示。

源文件 8-5　IStack.java

```
1  public interface IStack<E> {
2      /**
3       * 入栈
4       * @param e
5       *            入栈元素
6       * @throws RuntimeException.
7       */
8      void push(E e) throws RuntimeException;
9
10     /**
11      * 弹栈
12      * @return 栈顶元素
13      * @throws RuntimeException
14      */
```

```
15      E pop() throws RuntimeException;
16
17      /**
18       * 查看栈顶元素,并不改变栈的内容
19       * @return 栈顶元素
20       * @throws UnderflowException
21       */
22      E top() throws RuntimeException;
23
24      /**
25       * 判断栈空
26       * @return 如果栈空返回 true;否则返回 false
27       */
28      boolean isEmpty();
29
30      /**
31       * 判断栈满
32       * @return 如果栈满返回 true;否则返回 false
33       */
34      boolean isFull();
35
36      /**
37       * 清空
38       */
39      void clear();
40 }
```

源文件 8-5 中定义了 6 个抽象方法,分别完成入栈 push、弹栈 pop、查看栈顶元素 top、判断栈满 isFull、判断栈空 isEmpty 和清空 clear 操作。当由于容量限制而无法压栈等事件出现抛出运行时刻异常。当由于栈空而无法弹栈和查看栈顶元素时抛出运行时刻异常。

源文件 8-6 中定义的 MyStack<E> 是 IStack<E> 的一种基于定长容量线性表的实现。

源文件 8-6　MyStack.java

```
1 import   java.util.ArrayList;
2 /**
3  * 基于线性表的实现
4  * @author Donald Dong
5  */
6
7 public   class MyStack<E> implements IStack<E> {
8      private ArrayList<E> elements;
9      private int topOfStack;
10     private static final int DEFAULT_CAPACITY = 128;
11
12     public MyStack() {
13         elements =   new ArrayList<E>(DEFAULT_CAPACITY);
14         topOfStack = -1;
15     }
16
17     public void push(E e) throws RuntimeException {
18         if (this.isFull())
19             throw new RuntimeException("ArrayStack is already full.");
```

```
20        else {
21            elements.add(++topOfStack, e);
22        }
23    }
24
25    public E pop() throws RuntimeException {
26        if (this.isEmpty())
27            throw new RuntimeException("ArrayStack is already empty.");
28        else {
29            return (E) elements.get(topOfStack--);
30        }
31    }
32
33    public E top() throws RuntimeException {
34        if (this.isEmpty())
35            throw new RuntimeException("ArrayStack is empty now.");
36        else
37            return elements.get(topOfStack);
38    }
39
40    public boolean isEmpty() {
41        return topOfStack == -1;
42    }
43
44    public boolean isFull() {
45        return (topOfStack + 1 == DEFAULT_CAPACITY);
46    }
47
48    public void clear() {
49        topOfStack = -1;
50    }
51 }
```

源文件 8-7 演示了泛型接口 IStack 及其实现类泛型栈 MyStack 的使用。第 3 行创建了元素为字符串的栈，然后在第 4 行和第 5 行把 abc 和 xyz 两个字符串压栈；第 6 行和第 7 行弹栈，依次弹出 xyz 和 abc。第 8 行判断当前栈是否为空，输出结果为 true。第 10 行创建了元素为 Integer 类型的栈，第 11 行把整数 23 压栈；第 12 行查看栈顶元素，输出结果为 23。第 13 行弹栈，第 14 行再弹栈，由于栈已经空，抛出运行时刻异常：

```
Exception in thread "main" java.lang.RuntimeException: ArrayStack is already empty.
```

这个异常是在第 27 行抛出的。

源文件 8-7　MyStackTest.java

```
1 public class MyStackTest {
2     public static void main(String[] args) {
3         IStack<String> s = new MyStack<String>();
4         s.push("abc");
5         s.push("xyz");
6         System.out.println(s.pop());
7         System.out.println(s.pop());
8         System.out.println(s.isEmpty());
```

```
9
10      IStack<Integer> t = new MyStack<>();
11      t.push(23);
12      System.out.println(t.top());
13      t.pop();
14      t.pop();
15   }
16 }
```

程序的输出为：

```
xyz
abc
true
23
Exception in thread "main" java.lang.RuntimeException: ArrayStack is already empty.
    at MyStack.pop(MyStack.java:27)
    at MyStackTest.main(MyStackTest.java:14)
```

8.4 泛型方法

泛型类，是在实例化类的时候指明类型参数的具体类型；泛型方法则是在调用方法的时候指明类型参数的具体类型。

源文件 8-8 在第 6 行声明了一个泛型方法 getElement。该方法的第一个形式参数的类型是：

```
List<? extends Number>
```

意思是线性表的元素的类型必须是 Number 的后代类（含 Number）。通配符？表示任意类型，List<Integer> 与 List<? extends Number> 类型兼容。

然后源文件 8-8 试图调用泛型方法从一个整型线性表中查找首个元素；再从另外一个字符串线性表中查找首个元素。但是编译源文件 8-8，会出现编译错误：

```
GenericMethodDemo.java:17: error: incompatible types:
List<String> cannot be converted to List<? extends Number>
        System.out.println(getElement(b, 1));
                                      ^
```

意思是在第 17 行出现不兼容类型：不能把 List<String> 转换为 List<? extends Number>。这是因为 String 不是 Number 的后代类。

源文件 8-8　GenericMethodDemo.java

```
1 import java.util.ArrayList;
2 import java.util.List;
3
4 public class GenericMethodDemo {
5     //设定类型实参必须是 Number 类型或者 Number 类型的子类
6     static Number getElement(List<? extends Number> list, int index) {
7         return list.get(index);
8     }
```

```
9
10    public static void main(String[] args) {
11        List<Integer> a = new ArrayList<>();
12        a.add(23);
13        a.add(56);
14        System.out.println(getElement(a, 1));
15        List<String> b = new ArrayList<>();
16        b.add("xyz");
17        System.out.println(getElement(b, 1));
18    }
19 }
```

删除产生编译错误的第 17 行,那么程序的输出为:

```
56
```

约定使用一个大写字母作为类型形式参数的名字,常见的名字有:
<E>用于表示容器中元素的类型;
<T>用于表示类型;
<K,V>用于表示键(key)类型和值(value)类型;
<N>用于表示数值类型。

如果需要多个类型形式参数,则依次使用 S、U、V 等。表 8-1 以 java.util.List 类为例汇总了泛型定义中常见的类型形式参数的名字及其含义。

表 8-1 泛型中的形式参数

名 字	含 义
List<T>	元素类型为 T 的 List 类型
List<?>	任意元素类型的 List 类型(? 通配任意类型)
List<? super T>	元素类型是 T 的祖先类的 List 类型
List<? extends T>	元素类型是 T 的后代类的 List 类型
List<U extends T>	元素类型是 T 的直接子类的 List 类型

注意:List<? extends Number>类型的集体接收 Number 类型及其后代类型的元素,如 Integer 类型的对象或者 Double 类型的对象。但是 List<Integer>和 List<Number>却不是兼容类型。

下面的集体对象创建是错误的:

```
List<Number>  list  = new ArrayList<Integer>();
```

另外,在第 11 行 List<Integer> a=new ArrayList<>();中,ArrayList 后的<>中省略了 Integer。这是因为编译器通过类型推理能够知道 ArrayList 后的<>中是 Integer。

可在泛型方法头部返回值类型之前声明类型参数。例如:

```
public <T> List<T> toListFromArray(T[] a) {
    return Arrays.stream(a).collect(Collectors.toList());
}
```

把一个类型为 T 的数组转换为线性表。方法头部中的<T>表示该方法将处理泛型 T,这个声明是必需的,即使方法的返回值类型为 void。

源文件 8-9 定义并使用泛型方法 toListFromArray 把整数类型的数组{1,2,3,4,5}转换为字符串类型的线性表:("1","2","3","4","5")。

源文件 8-9   TestGenericM.java

```java
import java.util.List;
import java.util.Arrays;
import java.util.stream.Collectors;

public class TestGenericM {
    static <T> List<String> toListFromArray(T[] a) {
        return Arrays.stream(a)
            .map(Object::toString)
            .collect(Collectors.toList());
    }
    public static void main(String[] args) {
        Integer[] intArray = {1, 2, 3, 4, 5};
        List<String> stringList = toListFromArray(intArray);
        System.out.println(stringList);
    }
}
```

源文件 8-9 中的 static <T> List<String> toListFromArray(T[] a)是一个泛型方法,可以接受任何类型的数组,,例如 Integer[]、Double[]等,并将其转换为一个 String 列表。T 是泛型参数。Arrays.stream(a)将数组 a 转换为一个流;map(Object::toString)则把流中的每个元素转换为字符串。其中,Object::toString 是一个方法引用,表示调用每个对象的 toString 方法;collect(Collectors.toList())把流中的元素收集到一个字符串线性表 List<String> 中。

方法引用是一种简洁的语法,用于直接引用已有的方法或构造函数,而无需显式地创建 lambda 表达式。方法引用提高了源代码的可读性和简洁性。方法引用主要有四种类型:

(1) 引用静态方法,形如<类名>::<静态方法名>;

(2) 引用特定对象的实例方法,形如<实例>::<实例方法>;

(3) 引用任意对象的实例方法,形如<类名>::<实例方法>,流中的每个元素必须是该类的实例;

(4) 引用构造方法,形如<类名>::new。

第 8 章　章节测验

第9章 反 射

反射（reflection）就是 Java 程序在运行时刻获取类的信息的机制。对于任意一个类，Java 程序在运行状态中都能够知道这个类的所有属性和方法；对于任意一个对象，都能够调用它的任意一个方法和属性，这种动态获取类的信息以及动态调用对象的方法的机制称为 Java 语言的反射技术或反射机制。

通过反射，程序能够在运行时刻获得类的元数据（如类名、方法名、字段名）以创建对象、调用方法、访问和修改字段值等，而无须在编译时刻使用预知的类型创建和访问对象。Java 反射主要用于运行时动态加载类、创建对象和调用方法，广泛应用于框架设计。例如 JUnit 使用反射来解析@Test 注解从而得到测试方法然后调用该方法；Eclipse 使用反射实现方法名字的自动补全等。

一般情况下，写 Java 程序都是先设计类，再使用类声明、创建和访问对象。这些活动都是在程序运行之前完成的。尤其是程序员在使用类创建对象之前，往往对类有一定了解，知道类有哪些属性和方法等。有时需要在程序运行时刻根据需要动态加载某些类，而这些类的信息程序员预先未知。这就需要一套 API 帮助程序在运行时刻获取类的声明信息。

9.1 Class 类

类 java.lang.Class 是 Java 反射机制的一个非常重要的类。类 Class 的实例是 Java 进程中的类、接口、数组、基本数据类型、void 等类型。

假设已经在 Car.java 中声明了类 Car，如源文件 9-1 所示。类中声明了两个私有的成员变量：车牌号码 licenceNumber 和车身质量 weight、一个有参构造方法和一个无参构造方法、getter 和 setter 方法，以及覆盖方法 toString。

源文件 9-1 Car.java

```
1 public class Car {
2     private String licenceNumber;
3     private int weight;
4
5     public Car(String licenceNumber, int weight) {
6         this.licenceNumber = licenceNumber;
7         this.weight = weight;
8     }
9
10    public Car() {
11        this.licenceNumber = "ABCD";
12        this.weight = 1234;
13    }
14    public String getLicenceNumber() {
15        return licenceNumber;
16    }
17
18    public void setLicenceNumber(String licenceNumber) {
19        this.licenceNumber = licenceNumber;
20    }
21
22    public int getWeight() {
23        return weight;
24    }
25
26    public void setWeight(int weight) {
27        this.weight = weight;
28    }
29
30    public String toString() {
31        return "[" + licenceNumber + ", " + weight + "]";
32    }
33 }
```

类 Class 是一个泛型类：Class<T>。源文件 9-2 试图在程序运行时刻动态获取类名。第 3 行到第 5 行声明了三个引用变量准备引用 Class<?> 的实例，也就是任意类型的 Class 实例。对象的实例方法 getClass() 返回一个 Class 类型引用，用以引用 Class 类型的对象（此时是 Car）。

第 8 行通过类方法 Class.forName("Car") 动态装入 Car 类，并通过变量 a 引用这个 Class 实例，然后第 9 行调用 Class 类的实例方法 getName 返回 Class 实例的名字，也就是 Car。

第 11 行通过调用 Car 对象的实例方法 getClass() 获得 Class 实例，然后在第 12 行输出这个实例的名字，也就是 Car。

第 14 行通过访问类型的属性 class 得到 Class 实例。第 15 行的输出仍然是 Car。

源文件 9-2 ReflectClassDemo.java

```
1
2 public class ReflectClassDemo {
3     public static void main(String[] args) throws ClassNotFoundException {
```

```
4       Class<?> a = null;
5       Class<?> b = null;
6       Class<?> c = null;
7
8       a = Class.forName("Car");
9       System.out.println(a.getName());
10
11      b = new Car().getClass();
12      System.out.println(b.getName());
13
14      c = Car.class;
15      System.out.println(c.getName());
16    }
17 }
```

源文件 9-2 展示了在运行时刻获得 Class 实例的三种方式：类方法 Class.forName、实例方法 getClass 和类型的属性 class。

注意：在第 3 行声明了抛出异常 ClassNotFoundException，这个异常可能会在第 8 行产生。

9.2 实例化对象

源文件 9-3 展示了如何动态加载类后通过查询和调用构造方法实例化对象。在第 5 行动态加载了类 Car 并通过引用变量 c 引用这个 Class 实例。然后在第 8 行通过 Class 的实例方法 getDeclaredConstructor 查询类 Car 中声明的无参构造方法并通过实例方法 newInstance() 调用无参构造方法实例化对象。第 11 行通过 Class 的实例方法 getDeclaredConstructor 查询类 Car 中声明的有参构造方法 Car(String licenceNumber, int weight) 并通过实例方法 newInstance() 调用该有参构造方法实例化对象，注意此处使用了 .class 获取构造方法参数的类型而没有直接使用类型名字：getDeclaredConstructor(String.class, int.class)。

源文件 9-3　ReflectNewInstanceDemo.java

```
1  public class ReflectNewInstanceDemo {
2    public static void main(String[] args)
3        throws ReflectiveOperationException   {
4      Class<?> c = null;
5      c = Class.forName("Car");
6
7      Car car = null;
8      car = (Car) c.getDeclaredConstructor().newInstance();
9      System.out.println(car);
10
11     car = (Car) c.getDeclaredConstructor(String.class, int.class)
12         .newInstance("XYZ", 5678);
13     System.out.println(car);
14   }
15 }
```

程序的输出为：

```
[ABCD, 1234]
[XYZ, 5678]
```

注意：Class.forName() 可能会抛出 ClassNotFoundException 异常。这是受检异常，必须捕获或继续抛出；getDeclaredConstructor() 方法调用也可能会抛出 NoSuchMethodException 异常；newInstance() 方法调用可能会抛出 InstantiationException 异常，或者 IllegalAccessException 异常，或者 IllegalArgumentException，或者 InvocationTargetException。这些异常都是受检异常，需要捕获或继续抛出。ReflectiveOperationException 是这些异常的祖先类，第 3 行使用保留字 throws 声明继续抛出 ReflectiveOperationException 异常。

9.3 查询类的成员

使用 Class 的实例方法查询类的成员，查询结果是各种成员对象。包括 Field 类型的成员变量、Method 类型的成员方法，还有 Constructor 类型的构造方法。这些类型都在 java.lang.reflect 包中。

Class 的实例方法 getConstructors 以数组返回类的所有构造方法，实例方法 getMethods 以数组返回类的所有成员方法，实例方法 getDeclaredFields 以数组返回类的所有成员变量。源文件 9-4 展示了如何使用 Class 实例在运行时刻读取类 Car 的成员。

第 10 行首先使用类方法 Class.forName 动态装入类 Car。第 11 行调用 Class 的实例方法 getConstructors 查询 Car 类中声明的两个构造方法，并在第 11 行和第 12 行分别输出。第 16 行查询构造方法 Car(java.lang.String,int) 的参数类型，并在第 17 行输出：[class java.lang.String, int]。

第 19 行到第 21 行查询和输出在 Car 中声明的两个成员变量：

```
private java.lang.String Car.licenceNumber
private int Car.weight
```

第 23 行查询 Car 中声明的全部成员方法，第 24 行通过类方法 Arrays.toString 输出：

```
[public void Car.setWeight(int),
public int Car.getWeight(),
public java.lang.String Car.getLicenceNumber(),
public void Car.setLicenceNumber(java.lang.String),
public java.lang.String Car.toString()]
```

从第 26 行开始，程序试图"拼"出方法的头部。对于查询到的每个方法，依次查询该方法的修饰符并输出、查询方法的返回值类型并输出、查询方法名字并输出、查询方法的参数类型并输出参数列表，以及方法抛出的全部异常。

源文件 9-4 ReflectGetMembersDemo.java

```
1 import java.lang.reflect.Constructor;
2 import java.lang.reflect.Field;
3 import java.lang.reflect.Method;
4 import java.lang.reflect.Modifier;
5 import java.util.Arrays;
```

```java
6
7  public class ReflectGetMembersDemo {
8      public static void main(String[] args)
9              throws ReflectiveOperationException {
10         Class<?> c = Class.forName("Car");
11         Constructor<?>[] constructors = c.getConstructors();
12         System.out.println(constructors[0]);
13         System.out.println(constructors[1]);
14
15         Class<?>[] constructorParameterTypes
16             = constructors[0].getParameterTypes();
17         System.out.println(Arrays.toString(constructorParameterTypes));
18
19         Field[] fields = c.getDeclaredFields();
20         System.out.println(fields[0]);
21         System.out.println(fields[1]);
22
23         Method[] methods = c.getDeclaredMethods();
24         System.out.println(Arrays.toString(methods));
25
26         for (Method m : methods) {
27             //查询方法的修饰符
28             System.out.print(Modifier.toString(m.getModifiers()) + " ");
29             //查询方法的返回值类型
30             Class<?> returnType = m.getReturnType();
31             System.out.print(returnType.getName() + " ");
32             //查询方法的名字
33             System.out.print(m.getName());
34             System.out.print("(");
35             //查询参数类型
36             Class<?>[] parameterTypes = m.getParameterTypes();
37             //循环输出方法的参数
38             for (int p = 0; p < parameterTypes.length; p++) {
39                 System.out.print(parameterTypes[p].getName()
40                         + " " + "arg" + p);
41                 if (p < parameterTypes.length - 1) {
42                     System.out.print(",");
43                 }
44             }
45             //查询方法抛出的全部异常
46             Class<?>[] ex = m.getExceptionTypes();
47             //判断是否有异常
48             if (ex.length > 0) {

49                 System.out.print(") throws ");
50             } else {
51                 System.out.print(") ");
52             }
53             //输出异常信息
```

```
54          for (int j = 0; j < ex.length; j++) {
55              System.out.print(ex[j].getName());
56              if (j < ex.length - 1) {
57                  System.out.print(",");
58              }
59          }
60          System.out.println();
61      }
62  }
63 }
```

程序完整的输出为:

```
public Car(java.lang.String,int)
public Car()
[class java.lang.String, int]
private java.lang.String Car.licenceNumber
private int Car.weight
[public void Car.setWeight(int), public int Car.getWeight(), public java.lang.
String Car.getLicenceNumber(), public void Car.setLicenceNumber(java.lang.
String), public java.lang.String Car.toString()]
public void setWeight(int arg0)
public int getWeight()
public java.lang.String getLicenceNumber()
public void setLicenceNumber(java.lang.String arg0)
public java.lang.String toString()
```

利用 Class 类动态载入一个类之后,还可以查询该类实现的接口。例如源文件 9-5 在动态装入类 java.util.ArrayDeque 后在第 5 行调用实例方法 getInterfaces 查询 ArrayDeque 实现的所有接口。

源文件 9-5　ReflectGetInterfaces.java

```
1 public class ReflectGetInterfaces {
2     public static void main(String[] args)
3         throws ReflectiveOperationException{
4         Class<?> d = Class.forName("java.util.ArrayDeque");
5         Class<?>[] interfaces = d.getInterfaces();
6         for (int i = 0; i < interfaces.length; i++) {
7             System.out.println(interfaces[i].getName());
8         }
9     }
10 }
```

程序的输出为:

```
java.util.Deque
java.lang.Cloneable
java.io.Serializable
```

类似地,通过调用 Class 的实例方法 getSuperclass 查询父类,调用实例方法 getPackage 查询类从属的包,调用实例方法 getModifiers 查询类的修饰符,以及调用实例方法 getName 查询类的名字。

9.4 调用成员方法

首先利用 Class 的实例方法 getMethod 根据方法签名返回一个 Method 对象 method，然后调用 Method 对象 method 上的实例方法 invoke：

public Object invoke(Object obj, Object ... args)

意思是让对象 obj 使用参数 args 执行 method，返回 Object 类型的对象。源文件 9-6 试图动态装载 java.util.ArrayList 并调用其 add 方法、size 方法和 get 方法。在第 6 行动态装载该类，第 7 行根据方法签名 add(Object) 调用实例方法 getMethod 返回 Method 对象 addMethod。第 9 行使用无参数的构造方法创建一个线性表对象 list，第 11 行和第 12 行分别使用实际参数字符串 ABC 和 XYZ 在 list 上调用实例方法 add。

类似地，第 16 行在 list 上调用实例方法 size；第 20 行和第 21 行分别在 list 上以实际参数 0 和 1 调用 get 方法。

源文件 9-6 ReflectCallMethodDemo.java

```
1  import java.lang.reflect.Method;
2
3  public class ReflectCallMethodDemo {
4      public static void main(String[] args) {
5          try {
6              Class<?> clazz = Class.forName("java.util.ArrayList");
7              Method addMethod = clazz.getMethod("add", Object.class);
8
9              Object list = clazz.getDeclaredConstructor().newInstance();
10             //Invoke the add() method on the instance
11             addMethod.invoke(list, "ABC");
12             addMethod.invoke(list, "XYZ");
13
14             Method sizeMethod = clazz.getMethod("size");
15             //Invoke the size() method and print the result
16             System.out.println("Size: " + sizeMethod.invoke(list));
17
18             Method getMethod = clazz.getMethod("get", int.class);
19             //Invoke the get() method and print the result
20             System.out.println("First: " + getMethod.invoke(list, 0));
21             System.out.println("Second : " + getMethod.invoke(list, 1));
22         } catch (Exception e) {
23             e.printStackTrace();
24         }
25     }
26 }
```

程序的输出为：

```
Size: 2
First: ABC
Second: XYZ
```

第 9 章 章节测验

第 10 章 输入输出流

计算机的硬件系统由主机和外部设备组成。主机包含中央处理器内存和外存,外部设备分为输入设备和输出设备两类。"输入"指的是从外部设备,如磁盘、键盘等,把数据读入主机内部;"输出"指的是从主机把数据写入外部设备,如磁盘、显示器等;"流"是负责输入或输出的专用对象。Java 的流类似于快递公司,程序通过输入输出流从磁盘等外部设备得到自己期望的数据,类似于通过快递公司得到一本网购的图书。

10.1 文件与文件夹管理

文件系统是操作系统的组成部分之一。文件系统用以管理在持久化存储设备(硬盘、U 盘、固态盘)上存储的文件。如建立文件、删除文件、建立文件夹、删除文件夹、移动文件夹等。

通常把文件组织在不同的文件夹(folder)中,文件夹是 Windows 操作系统中的术语,在 Linux 等操作系统中称为目录(directory)。文件夹中既可以包含文件,也可以有文件夹,文件夹中的文件夹称为子文件夹。直接包含文件的文件夹称为该文件的父文件夹。所以,文件夹形成了树状层次结构。从树的根开始的文件夹、子文件夹及文件称为卷(volume)。一般在一个磁盘上建立多个卷。

图 10-1 在 Windows 命令提示符窗口下使用 dir 命令显示了 C 盘 work 文件夹中的文件及文件夹。这个命令首先显示"驱动器 C 中的卷是 Windows,卷系列号是 8C4F-9DB4"。然后列表显示了文件的创建日期、时间、类型、大小和文件名。其中<DIR>表示该项是文件夹。例如 car 就是一个文件夹;Car.java 是个文件,其大小是 769 字节。

文件通过文件名来引用。文件名包括"."分隔的基本名和扩展名两部分。文件名中最后一个点"."之前的字符序列称为基本名(base name)。在文件名中最后一个点"."之后的字符序列称为扩展名(file name extension),通常用来标识文件的类型,如 Car.java 中的

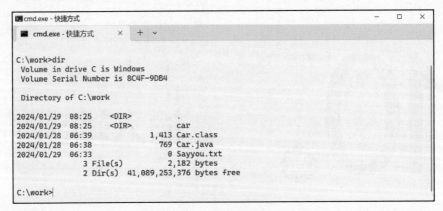

图 10-1　卷、目录和文件

java 就是扩展名,表示 Java 源文件;Sayyou.txt 中的扩展名 txt 表示文本文件。

到达某个文件的文件夹序列称为路径(path)。在 Windows 系统中,路径中的文件夹名使用反斜线"\"作为分隔符。例如 JDK 中的可执行文件 javac.exe 一般的安装路径是:

```
C:\Program Files\Java\jdk\bin\javac.exe
```

但是,在 Linux 操作系统中,路径中的文件夹却使用斜线"/"作为分隔符:

```
/usr/local/jdk /bin/javac.exe
```

有的操作系统用冒号":"作为路径中的分隔符。

从根开始到达某个文件的文件夹序列称为全路径(full path),也称为绝对路径(absolute path)。从当前文件夹开始到达某个文件的文件夹序列称为相对路径(relative path)。当前文件夹使用"."表示,父文件夹使用".."表示。

在 Java 程序中对磁盘文件进行的操作包括创建、删除、移动、重命名、查询等。java.io. File 类提供了与平台无关的对文件或文件夹的操作方法,如创建、重命名、复制、删除等。一个 File 对象是对一个磁盘文件或文件夹进行操作的"操作员",Java 程序通过 File 对象实现对文件的操作,正如网上书店的客户通过快递公司得到一本书一样。

通常使用如下构造方法创建 File 对象:

File(String) 以字符串表示的文件名或文件夹名作为参数创建对象。文件名或文件夹名前可缀以绝对路径,也可缀以相对路径。

File(String,String) 以字符串表示的父文件夹名和文件名作为参数创建对象。

当使用构造方法创建了 File 对象后,仅仅准备好了一个对象去访问指定的磁盘文件,但该文件未必已经在磁盘上存在。

java.io.File 提供了一套访问文件或文件夹的实例方法,比如创建、删除、重命名、移动等。

createNewFile 方法以参数指定的绝对文件名创建文件。当且仅当文件不存在且创建成功返回 true;如果文件已经存在则返回 false。如果遇到 IO 错误则抛出 IOException 异常。mkdir 方法以参数指定的绝对文件夹名创建文件夹。exists 方法判断当前 File 对象所映射的磁盘文件或文件夹是否存在。length 方法返回当前 File 对象所映射的磁盘文件大小,以字节为单位。renameTo 重命名方法把当前 File 文件改名为 File 类型参数所指定文件名。由于 File 对象一旦创建不可更改，所以只能重新建立 File 对象与磁盘文件映射才能实现改名。delete 方法试图立即删除当前的 File 对象所映射的磁盘文件或文件夹。deleteOnExit 方法试图在程序运行结束时删除当前的 File 对象所映射的磁盘文件或文件夹。getName 方法返回当前 File 对象映射的磁盘文件或文件夹的名字。getPath 方法返回当前 File 对象映射的磁盘文件或文件夹的绝对路径及文件名。isDirectory 方法判断 File 对象是否映射文件夹。list 方法以字符串数组形式返回当前文件夹中的内容。listFiles 方法以 File 对象数组形式返回当前文件夹中的内容。

源文件 10-1 演示了与操作磁盘文件有关的方法。该程序首先创建了访问磁盘文件 "C:\work\Sayyou.txt" 的 File 对象，然后通过该 File 对象的 createNewFile 方法在磁盘上创建文件。创建之前先判断该文件是否存在，如果已经存在则删除。

源文件 10-1　NewFileTest.java

```
1 import java.io.File;
2 import java.io.IOException;
3
4 public class NewFileTest {
5     public static void main(String[] args) throws IOException {
6         File f = new File("C:\work\Sayyou.txt");
7         if (f.exists()) {
8             f.delete();
9         }
10        f.createNewFile();
11        System.out.println(f.getName());
12        System.out.println(f.getPath());
13        System.out.println(f.isFile());
14        System.out.println(f.isDirectory());
15    }
16 }
```

注意：在源文件 10-1 中，File 构造方法的参数使用了转义符号"\"把转义符号"\"转义成普通字符。

如果没有异常抛出，程序第 11 行到第 14 行依次让 File 对象执行 getName、getPath、isFile、isDirectory 方法。输出分别是：

```
Sayyou.txt
C:\work\Sayyou.txt
true
false
```

其中，Sayyou.txt 是文件名；C:\work\Sayyou.txt 是文件的绝对路径名；true 是方法 isFile 的返回结果，表示是文件而不是文件夹；false 是方法 isDirectory 的返回结果，表示不是文件夹。

在源文件 10-1 中使用了 "\" 作为文件路径的分隔符，但这个分隔符是 MS Windows 文件系统所特定的。这种写法使得程序不具有可移植性。也就是说，能够在 MS Windows 上运行的程序却不能在 UNIX 上运行。File 类中定义了静态成员 separator 用以存储不同操作系统中所使用的分隔符：在 UNIX 中其值为"/"；在 MS Windows 中其值为"\"。

源文件 10-2 是源文件 10-1 的可移植版本。

源文件 10-2　NewFilePortable.java

```
1 import java.io.File;
2 import java.io.IOException;
3
4 public class NewFilePortable {
5     public static void main(String[] args) throws IOException {
6         File f = new File("c:" + File.separator
7             + "work" + File.separator + "Sayyou.txt");
8         if (f.exists()) {
9             f.delete();
10        }
11        f.createNewFile();
12        System.out.println(f.getName());
13        System.out.println(f.getPath());
14        System.out.println(f.isFile());
15        System.out.println(f.isDirectory());
16    }
17 }
```

在源文件 10-2 中，第 6 行使用相对路径和 File.separator 组成含文件路径前缀的文件名："."＋File.separator＋"work"＋File.separator＋"Sayyou.txt"，并以此作为参数创建 File 对象。在 MS Windows 中即试图创建".\work\Sayyou.txt"。

注意，运行程序前应确认"C:\work"文件夹已经存在。如果该文件夹不存在，那么程序会抛出异常。若"C:\work"文件夹已经存在，程序的输出是：

```
Sayyou.txt
C:\work\Sayyou.txt
true
false
```

如果在 Windows 资源管理器中打开"C:\work"文件夹，就会发现该文件夹中有一个 0 字节的文本文件 Sayyou.txt。

源文件 10-3 演示了文件夹的创建和访问。

源文件 10-3　NewFolderTest.java

```
1 import java.io.File;
2
3 public class NewFolderTest {
4     public static void main(String[] args) {
5         File parent = new File("C:\\java");
6         parent.mkdir();
7         File dir = new File("C:\\java\\work");
8         dir.mkdir();
9         String[] content = parent.list();
10        for (String item : content) {
```

```
11                  System.out.println(item);
12              }
13      }
14 }
```

在源文件 10-3 中,第 5 行创建 File 对象映射文件夹"C:\java",第 6 行让该 File 对象执行 mkdir 方法创建文件夹"C:\java"。第 7 行创建了 File 对象映射"C:\java\work"文件夹。第 8 行让 File 对象执行 mkdir 方法在磁盘上创建文件夹"C:\java\work"。第 9～12 行列出文件夹"C:\java"中的内容:File 对象 parent 执行 list 方法以字符串数组返回文件夹中的内容。最后使用 for-each 遍历这个数组。输出结果中只有一项,即刚创建的文件夹:

```
work
```

源文件 10-4 演示了如何通过实例方法访问文件的属性。

源文件 10-4　AttributesTest.java

```
1 import java.io.File;
2
3 public class AttributesTest {
4      public static void main(String[] args) {
5          File myFile = new File("C:\\work\\Sayyou.txt");
6          System.out.println(myFile
7              + (myFile.exists() ? " exists" : " does not  exist"));
8          System.out.println(myFile + " is a "
9              + (myFile.isFile() ? " FILE" : " FOLDER"));
10         System.out.println("The length of " + myFile + " is "
11             + myFile.length() + " bytes");
12         System.out.println(myFile
13             + (myFile.isHidden() ? " is" : " is not") + " hidden");
14         System.out.println("You can "
15             + (myFile.canRead() ? " " : "not ") + "read " + myFile);
16         System.out.println("You can "
17             + (myFile.canWrite() ? " " : "not ") + "write " + myFile);
18     }
19 }
```

在源文件 10-4 中,第 5 行创建 File 对象用以访问已经在磁盘上创建好的文本文件"C:\work\Sayyou.txt"。第 7 行让 File 对象执行 exists 方法判断该对象映射的磁盘文件 Sayyou.txt 是否存在。第 9 行调用实例方法 isFile 判断是文件还是文件夹。第 11 行调用实例方法 length 返回以字节为单位的文件长度。第 13 行调用实例方法 isHidden 判断是否是隐藏文件。第 15 行和第 17 行通过执行 canRead 方法和 canWrite 方法判断 Sayyou.txt 是否允许读和允许写。程序的输出为:

```
C:\work\Sayyou.txt exists
C:\work\Sayyou.txt is a FILE
The length of C:\work\Sayyou.txt is 0 bytes
C:\work\Sayyou.txt is not hidden
You can read C:\work\Sayyou.txt
You can write C:\work\Sayyou.txt
```

由于文件夹中既允许包含文件,又可以包含文件夹,如果想列出文件夹及其子文件夹中的内容,则需遍历文件夹及其子文件夹。源文件 10-5 以递归方式进行文件夹的遍历。

源文件 10-5　ListFolderTest.java

```java
1  import java.io.File;
2
3  public class ListFolderTest {
4      public static void main(String[] args) {
5          File dir = new File("c:\\work");
6          travel(dir);
7      }
8
9      public static void travel(File dir) {
10         if (dir.isDirectory()) {
11             File[] items = dir.listFiles();
12             for (File item : items) {
13                 System.out.println(item.getAbsoluteFile());
14                 if (item.isDirectory())
15                     travel(item);
16             }
17         }
18     }
19 }
```

在源文件 10-5 的程序中第 6 行调用静态方法 travel 进行深度优先文件夹遍历。方法 travel 首先判断参数指定的 File 对象是否是文件夹。如果是文件夹,则让该 File 对象执行方法 listFiles 返回文件夹中的所有项目,然后使用 for-each 循环逐一遍历这些项目:首先输出项目的绝对路径名;然后判断该项目是否有文件夹,是则以该项目作为参数,递归调用 travel 方法。

假设已经有磁盘文件夹 C:\work,其中有文件夹 car、文件 Car.java、文件 Car.class、文件 Sayyou.txt,如图 10-2 所示。文件夹 car 中还有文件夹 cs,cs 中还有文件夹 dd,文件夹 dd 中只有一个文件 Car.class,如图 10-3 所示。

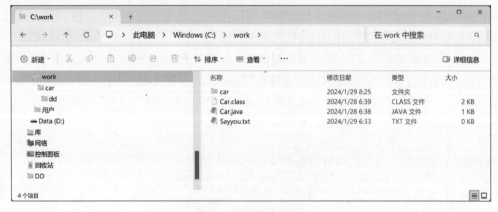

图 10-2　文件夹 work

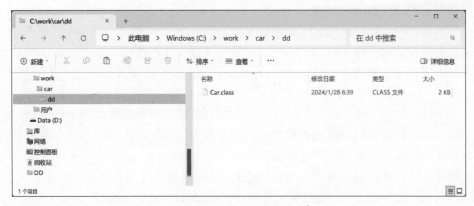

图 10-3 文件夹 dd

那么，源文件 10-5 中的程序输出是：

```
c:\work\car
c:\work\car\dd
c:\work\car\dd\Car.class
c:\work\Car.class
c:\work\Car.java
c:\work\Sayyou.txt
```

 流

商品从库房运出，最终交付给用户，就形成了物流。数据从本地磁盘或者远程服务器读出，最终交付给本地进程，就形成了数据流。进程可以从键盘、磁盘、网络、内存缓冲区或者其他进程数据读取输入，这些位置称为数据的来源（source）。进程可以把数据输出到显示器、磁盘、网络、内存缓冲区或者另外的进程，这些位置称为数据的归宿（sink）。

Java 的输入流（input stream）对象负责从数据的来源读取数据；输出流（output stream）对象负责把数据写入数据的归宿中。需要读写数据的 Java 程序通过创建输入流对象和输出流对象完成数据的读写。图 10-4 展示了输入流的概念。

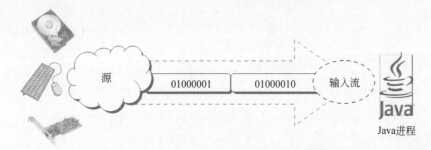

图 10-4 输入流

从 Java 程序角度看，输入流就像负责把数据从指定位置搬运到本地变量中的快递公司；输出流就像把本地变量中的数据搬运到指定位置的快递公司。流封装了与数据源或者

数据归宿有关的访问细节，使得客户程序仅关心数据本身。流的客户程序不必关心具体设备的读写机制，即使设备发生变化，如磁盘变成了固态盘，客户程序也无须做任何修改。

流就是无存储的比特序列，即 0 或 1 的序列。但是，计算机按字节存储和访问数据，所以流的基本元素是字节或者字符。这样，流分为两种类型：字节流（byte stream）和字符流（character stream）。字节流把比特序列看作字节序列；而字符流把比特序列看作是字符序列。一个字符编码为 1 字节或若干字节。必须按照指定的编码规则访问字符流，比如如果以 UTF-8 编码规则访问 GBK 编码的字符就会显示为"乱码"。以字符方式存储的数据源，如文本文件、网页，都应使用字符流对象读写。

使用输入流读数据的步骤有三个：①创建输入流对象并将其与数据来源相关联；②让输入流对象执行读操作；③当数据输入完毕，输入流对象执行关闭操作。

使用输出流写数据的步骤也有三个：①创建输出流对象并将其与数据归宿相关联；②让输出流对象执行写操作；③当数据输出完毕，输出流对象执行关闭操作。

10.2.1　字节流

java.io.InputStream 类和 java.io.OutputStream 类用于完成字节流的输入和输出。但是，这两个类是抽象类，不能创建其类型的对象。但可以使用其后代类创建对象。比如以字节为单位读磁盘文件的 FileInputStream，按照基本数据类型读字节流的 DataInputStream 等。OutputStream 类的后代类有以字节为单位写磁盘文件的 FileOutputStream、按照基本数据类型写字节流的 DataOutputStream 等。

可以把文件名作为字符串参数使用 FileOutputStream(String) 构造方法创建对象，也可以先创建 File 对象，以 File 对象作为参数使用 FileOutputStream(File) 构造方法创建对象。如果参数指定的文件已经存在，则原文件被清除。如果希望保留原文件的内容，向其中追加数据，则在调用构造方法时增加一个参数 true 指示进行追加。即使用构造方法 FileOutputStream(String，boolean) 或 FileOutputStream(File，boolean)。源文件 10-6 演示了如何使用字节流写文件。

源文件 10-6　ByteFileOutputTest.java

```java
1 import java.io.FileOutputStream;
2 import java.io.File;
3 import java.io.IOException;
4
5 public class ByteFileOutputTest {
6     public static void main(String[] args) {
7         byte[] data = { 1, 2, 3, 4, 5, 0, 6, 7, 8, 9, 10, 0 };
8         File outFile = new File("out.dat");
9         try (FileOutputStream fout = new FileOutputStream(outFile)) {
10             fout.write(data);
11             System.out.println("Total bytes: " + outFile.length());
12         } catch (IOException e) {
13             e.printStackTrace();
14         }
15     }
16 }
```

在源文件 10-6 中，第 7 行将准备写入文件的 12 个整数放在了字节数组 data 中。第 8

行创建了 File 类型的对象用于访问到当前文件夹下 out.dat 文件。在第 9 行到开始的 try 块中是完成数据输出的语句序列：让输出流对象 fout 执行 write 方法输出字节数组 data 中的所有字节；让 File 对象 outFile 执行 length 方法以字节为单位返回文件的长度。注意，创建输出流对象 fout 的语句在保留字 try 后面的圆括号中，这使得在程序中无须显式地使用 finally 块让输出流对象 fout 执行 close 方法关闭输出流。程序的输出结果是：

```
Total bytes: 12
```

从输出结果看，文件中有 12 字节，这与字节数组中有 12 个整数一致。注意，目标文件 out.dat 将建立在当前文件夹中。如果使用 Eclipse，则会创建在项目文件夹中。

源文件 10-7 演示了如何以字节为单位从文件中把数据读入字节数组中。

源文件 10-7　ByteFileInputToArrayTest.java

```java
1  import java.io.File;
2  import java.io.FileInputStream;
3  import java.io.IOException;
4  import java.util.Arrays;
5
6  public class ByteFileInputToArrayTest {
7      public static void main(String[] args) {
8          File inFile = new File("out.dat");
9          byte[] data = new byte[16];
10
11         int n = 0;
12         try (FileInputStream fin = new FileInputStream(inFile)) {
13             n = fin.read(data);
14         } catch (IOException e) {
15             e.printStackTrace();
16         }
17
18         System.out.println("Total bytes read: " + n);
19         System.out.print(Arrays.toString(data));
20     }
21 }
```

在源文件 10-7 中，首先在第 8 行定义了 File 对象访问在源文件 10-6 中已经写入了 12 个数据的 out.dat 文件。第 9 行声明一个具有 16 个元素的字节数组 data 准备存放从文件中读取的数据。第 12 行到第 16 行的 try 块中是完成读入的语句序列：让 FileInputStream 对象 fin 执行 read 方法把数据以字节为单位读入字节数组 data 中。如果文件中的数据多于字节数组的容量(16 个)，则仅读入 16 字节；如果文件中的数据比字节数组的容量小，即不够 16 字节，则读到文件尾为止。read 方法返回实际读入的字节数。注意，创建 FileInputStream 对象 fin 用来访问文件 out.dat 的语句写在了 try 保留字后面的圆括号中，这称为 try-with-resources(带资源的 try 块)。try-with-resources 使得程序略去了显式的文件关闭语句。程序最后输出实际读入的字节数和完成读入后字节数组的内容。

通过上面的两个例子看到，一般需要把进行输入输出操作的语句序列放在 try 块中。这是因为输入输出操作都可能会抛出 IOException 异常。前文讲过，IOException 是 "checked"异常，必须在程序中声明处理方式：捕获还是继续抛出。源文件 10-6 和源文件 10-7 都是捕获处理方式。

如果把输入流或输出流的关闭操作放在了 try 块中，那么当执行读操作或写操作期间抛出异常，就会使得流的关闭操作无法执行。由于无论正常读写完毕还是出现异常都有执行流的关闭操作，Java 7 以前的程序一般把流的关闭操作放在 finally 块中。源文件 10-8 演示了如何在 finally 块中让流对象执行关闭操作。

源文件 10-8　ByteFileInputTest.java

```java
1  import java.io.File;
2  import java.io.FileInputStream;
3  import java.io.IOException;
4  
5  public class ByteFileInputTest {
6  
7      public static void main(String[] args) {
8          File inFile = new File("out.dat");
9          int i;
10         FileInputStream fin = null;
11         try {
12             fin = new FileInputStream(inFile);
13             i = fin.read();
14             while (i != -1) {
15                 System.out.print(i);
16                 i = fin.read();
17             }
18         } catch (IOException e) {
19             System.out.println(e.getMessage());
20         } finally {
21             try {
22                 fin.close();
23             } catch (Exception e) {
24                 System.out.println(e.getMessage());
25                 return;
26             }
27         }
28     }
29 }
```

在源文件 10-8 中，关闭文件的 close 方法安排在了 finally 块中。但由于该方法也可能抛出 IOException 异常，所以仍然需要把该方法嵌入在 try-catch 中。另外，这里让输入流对象直接执行 read 方法读取文件返回下一字节，而没有指定存放数据的字节数组。注意，当到达文件尾时，read 方法返回 −1。还需要注意，虽然每次读取一字节，但是 read 方法的返回值类型却是 int 而不是 byte。这是因为 −1 的 8 比特补码是 11111111，这是一字节所能表示的数之一，不管是有符号数 −1 还是无符号数 255。如果 read 方法的返回值是字节类型，就无法使用 −1 区分正常数值还是文件尾。通过在读取的 1 字节前补 3 字节，这 3 字节的每个比特都是 0，就能够解决这个问题。因为正常的数值前面都是 24 个 0，只有文件尾标识为 32 个 1。

如果 out.dat 文件不存在，那么程序的输出为：

```
out.dat (系统找不到指定的文件。)
Cannot invoke "java.io.FileInputStream.close()" because "fin" is null
```

在源文件 10-8 中的第 18 行的语句捕获了第 12 行语句抛出的异常，第 19 行的语句输出了"out.dat（系统找不到指定的文件。）"，接着执行 finally 块试图关闭流。但是，此时流 fin 为空，试图关闭流的语句 fin.close() 抛出异常。这个异常被第 23 行的语句捕获，第 24 行的语句输出 Cannot invoke "java.io.FileInputStream.close()" because "fin" is null。

如果把创建 FileInputStream 对象的语句放在 try 后面的括号中，则不用写显式的 finally 块关闭资源。源文件 10-9 是源文件 10-8 的重构版本。

源文件 10-9　TryWithResourceTest.java

```
1 import java.io.File;
2 import java.io.FileInputStream;
3 import java.io.IOException;
4
5 public class TryWithResourceTest {
6
7     public static void main(String[] args) {
8         int b;
9         try (FileInputStream fin
10             = new FileInputStream(new File("out.dat"))) {
11             b = fin.read();
12             while (b != -1) {
13                 System.out.print(b);
14                 b = fin.read();
15             }
16         } catch (IOException e) {
17             System.out.println(e.getMessage());
18         }
19     }
20 }
```

源文件 10-9 的第 9 行在 try 后面的圆括号中声明和创建了 FileInputStream 对象来访问磁盘文件，磁盘文件是一种资源（resource）。然后使用 catch 块捕获和处理异常或者在第 7 行声明抛出即可，不再用 finally 块释放资源。try-with-resources 语句保证了每个在 try 后面声明了的资源在 try 块结束的时候都会被关闭。但是注意，只有实现了 java.lang.AutoCloseable 接口的对象，或实现了 java.io.Closeable 接口的对象，才可以当作资源使用。

10.2.2　缓冲字节流

InputStream 对象的 read 方法的每次执行仅能读取 1 字节；OutputStream 对象的 write 方法每次执行仅能写入 1 字节。如果把 1 字节的数据看作是 1 粒米，那么把 1 字节的数据从内存输送磁盘或从磁盘传输到内存就好比是把 1 粒米从泰山脚下的泰安市运到泰山山顶。显然，如果按粒运输效率太低了。按字节读写同样效率太低了。如何提高效率呢？一种办法是把米装成麻袋，先把麻袋使用货车存放到泰山脚下的粮食仓库；然后再安排工人把麻袋扛到山顶。这里的"粮食仓库"就是缓冲区。同理，当试图把大量数据写入磁盘时，也需要建立缓冲区。

具有缓冲区的字节流 BufferedInputStream 和 BufferedOutputStream 就用来完成大量数据的读取和写入，简称为缓冲字节流。创建缓冲输入流的构造方法有两个：BufferedInputStream(InputStream) 和 BufferedInputStream(InputStream, int)。前者访问

参数指定的字节流对象,通过这个字节流对象把字节缓冲到自己的缓冲区中;后者则可以通过参数指定缓冲区的大小。

从缓冲输入流中读取数据最简单的方法是让缓冲输入流对象调用无参数的 read 方法。这个方法从缓冲输入流对象管理的缓冲区中返回 1 字节(0～255)。如果到达了输入流的末尾,没有字节可读取,则返回−1。

类似地,创建缓冲输出流的构造方法也有两个:BufferedOutputStream(OutputStream) 和 BufferedOutputStream(OutputStream,int)。前者访问参数指定的字节流对象,通过这个字节流对象把字节缓冲到自己的缓冲区中;后者则通过参数可以指定缓冲区的大小。把数据写入缓冲输出流中最简单的方法是让缓冲输入流对象调用 write(int) 方法。该方法把 1 字节(0～255)输送到缓冲输出流对象管理的缓冲区中。注意,write 方法的参数是 int 而不是 byte,写入时会将参数的高位截断。另外一个方法是 write(byte[ ], int, int)。这个方法从参数指定字节数组起始位置索引,写入指定字节数。

由于数据仅仅输出到了缓冲区,并没有实际输出到目标文件,所以还需让缓冲输出流对象执行 flush 方法以完成实际写入操作。当缓冲输出流对象执行 close 方法时会自动执行 flush 方法。

以复制图像文件 in.jpg 到 out.jpg 为例,源文件 10-10 演示了缓冲输入流和缓冲输出流的用法。in.jpg 的大小是 7003433 字节,约 7MB。

源文件 10-10　BufferedStreamDemo.java

```
1 import java.io.BufferedInputStream;
2 import java.io.BufferedOutputStream;
3 import java.io.File;
4 import java.io.FileInputStream;
5 import java.io.FileOutputStream;
6 import java.io.IOException;
7
8 public class BufferedStreamDemo {
9     public static void main(String[] args) {
10        String inFilePath = "in.jpg";
11        String outFilePath = "out.jpg";
12        long startTime = 0, elapsedTime = 0;
13
14        File fin = new File(inFilePath);
15        System.out.println("File size is " + fin.length() + " bytes");
16
17        try (BufferedInputStream in = new BufferedInputStream(new
18                FileInputStream(inFilePath));
19             BufferedOutputStream out = new BufferedOutputStream(new
20                FileOutputStream(outFilePath))) {
21
22            startTime = System.nanoTime();
23            int b;
24            while ((b = in.read()) != -1) {
25                out.write(b);
26            }
27        } catch (IOException e) {
28            e.printStackTrace();
```

```
29          }
30
31          elapsedTime = System.nanoTime() - startTime;
32          System.out.println("Elapsed Time is "
33              + (elapsedTime / 1000000.0)
34              + " msec");
35      }
36 }
```

在源文件 10-10 中,程序第 14 行声明通过 File 对象访问 in.jpg,第 15 行使用 File 对象的 length 方法返回 in.jpg 的文件大小,显示输出为:

```
File size is 7003433 bytes
```

然后第 17 行使用 try-with-resources 语句访问 File 对象的缓冲输入流和缓冲输出流对象。第 24 行 while 循环每次从文件 in.jpg 中读取 1 字节写入文件 out.jpg 中,直到文件结束(read 方法返回 -1)。

为了度量复制文件所消耗的时间,在开始复制之前,第 22 行使用静态方法 System. nanoTime 读取精确到纳秒的系统时间;在循环结束后,再使用该静态方法读取系统时间。第 31 行使用后者减去前者就是复制文件所消耗的时间。

程序第 33 行输出时把纳秒单位的时间除以 10^{-6},转换成毫秒为单位的时间。一个可能的输出结果为:

```
Elapsed Time is 137.6041 msec
```

运行程序的计算机不同,这个时间也会不同。

不使用缓冲流,而是直接使用字节流,那么复制文件所需时间大不相同,如源文件 10-11 所示。

源文件 10-11　FileCopyWithoutBuffer.java

```
1 import java.io.FileInputStream;
2 import java.io.FileOutputStream;
3 import java.io.IOException;
4
5 public class FileCopyWithoutBuffer {
6     public static void main(String[] args) {
7         String inFilePath = "in.jpg";
8         String outFilePath = "out.jpg";
9         long startTime = 0, elapsedTime = 0;
10
11        try (FileInputStream in = new
12                FileInputStream(inFilePath);
13            FileOutputStream out = new
14                FileOutputStream(outFilePath)) {
15            startTime = System.nanoTime();
16            int b;
17            while ((b = in.read()) != -1) {
18                out.write(b);
19            }
```

```
20          } catch (IOException e) {
21              e.printStackTrace();
22          }
23
24          elapsedTime = System.nanoTime() - startTime;
25          System.out.println("Elapsed Time is "
26              + (elapsedTime / 1000000.0)
27              + " msec");
28      }
29 }
```

源文件 10-11 与源文件 10-10 的不同在于第 16 行和第 18 行，这两行直接使用文件输入流每次读入 1 字节，使用文件输出流每次写入 1 字节。程序运行结果显示：

```
Elapsed Time is12424.8721 msec
```

可以看到，大约使用了 12 秒，远远慢于缓冲方式的复制。

10.2.3 数据流

如果程序需要从键盘、文件或者网络字节流中读取一个整数或者一个浮点数。这种情况适合使用 java.io.DataInputStream 或者 java.io.DataOutputStream 类。DataInputStream 对象使用 InputStream 对象的字节流输出，把字节流整理成 8 种基本数据类型的数据，如图 10-5 所示。

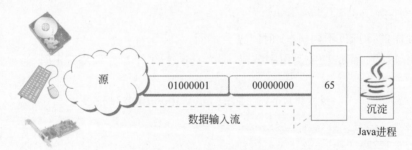

图 10-5　DataInputStream 过滤 InputStream 输出

在图 10-5 中，InputStream 对象的输出是字节序列"00000000""01000001"。在此基础上，如果客户程序要求输入 short 类型的整数，则 DataInputStream 对象将这 2 字节整合成一个 short 类型的整数：65。

DataOutputStream 以 8 种基本数据类型的数据流作为输入，把数据流解析成字节流，然后交字节流对象输出。比如，DataOutputStream 把一个 int 类型的值分解成连续的 4 字节。DataInputStream 和 DataOutputStream 都是过滤流（filtered stream）。

创建数据输入流的构造方法只有一个：DataInputStream(InputStream)，其参数是字节流。意思是按照给定的字节流创建数据输入流。InputStream 子类的对象，比如BufferedInputStream、FileInputStream 都可以作为构造方法的参数。

创建数据输出流的构造方法只有一个：DataOutputStream(OutputStream)，其参数是字节流。意思是按照给定的字节流创建数据输出流。OutputStream 子类的对象，比如BufferedOutputStream、FileOutputStream 都可以作为构造方法的参数。

表 10-1 数据输入流和数据输出流的方法签名列出了 DataInputSream 的 8 个读方法和 DataOutputStream 的 8 个写方法。

表 10-1 数据输入流和数据输出流的方法签名

DataInputStream	DataOutputStream	DataInputStream	DataOutputStream
readByte()	writeByte(int)	readFloat()	writeFloat(float)
readShort()	writeShort(int)	readDouble()	writeDouble(double)
readInt()	writeInt(int)	readChar()	writeChar(int v)
readLong()	writeLong(long)	readBoolean()	writeBoolean(boolean)

每个输入方法返回方法名所指示的数据类型的值。例如 readFloat 方法返回 float 类型的值。注意，writeChar 方法的参数是 int 类型，但实际只输出低位 2 字节。源文件 10-12 演示了 DataOutputStream 的用法。

源文件 10-12 DataOutputStreamDemo.java

```
1  import java.io.FileOutputStream;
2  import java.io.DataOutputStream;
3  import java.io.IOException;
4
5  public class DataOutputStreamDemo {
6      public static void main(String[] args) throws IOException {
7          FileOutputStream fo = new FileOutputStream("OutputFile.dat");
8          DataOutputStream dos = new DataOutputStream(fo);
9          dos.writeByte(-1);
10         dos.writeInt(16);
11         dos.writeLong(32);
12         dos.writeBoolean(true);//1 byte
13         dos.writeFloat(5.6f);
14         dos.writeDouble(3.14);
15         dos.writeChar('A');
16         System.out.println("Number of bytes written: " + dos.size());
17         dos.close();
18     }
19 }
```

源文件 10-12 第 7 行创建了 FileOutputStream 对象，准备通过该对象向磁盘文件 Output.dat 中输出。第 8 行创建数据过滤流对象 DataOutputStream 访问 FileOutputStream 对象。

第 9 行到第 22 行分别使用 writeXXX 方法以字节方式向磁盘文件中写入数据。例如执行 writeInt 方法写入整数 16；执行 writeDouble 方法写入 3.14。

第 16 行使用 size 方法返回当前流已经输出的字节个数：

Number of bytes written: 28

注意：所有的 writeXXX 方法都可能抛出异常，但是程序中并没有使用 try-catch 块捕获异常，而是在第 6 行的方法头部声明抛出异常。

源文件 10-13 演示了如何使用 DataInputStream 从文件中读取数据。

源文件 10-13　DataInputStreamDemo.java

```
1 import java.io.FileInputStream;
2 import java.io.DataInputStream;
3 import java.io.IOException;
4
5 public class DataInputStreamDemo {
6     public static void main(String[] args) throws IOException {
7         FileInputStream fo = new FileInputStream("OutputFile.dat");
8         DataInputStream dis = new DataInputStream(fo);
9         System.out.println(dis.readByte());
10        System.out.println(dis.readInt());
11        System.out.println(dis.readLong());
12        System.out.println(dis.readBoolean());
13        System.out.println(dis.readFloat());
14        System.out.println(dis.readDouble());
15        System.out.println(dis.readChar());
16        dis.close();
17    }
18 }
```

源文件 10-13 读取源文件 10-12 写入文件 OutputFile.dat 中的数据。程序第 7 行首先创建 FileInputStream 字节流对象用来访问磁盘文件 OutputFile.dat。然后创建 DataInputStream 对象用来以不同数据类型访问 FileInputStream 字节流对象。

第 9 行及其后续行使用 readXXX 方法从文件中读取不同类型的数据，并输出显示：

```
-1
16
32
true
5.6
3.14
A
```

如果试图从文件读取一个数据，而已经到达文件尾，那么 readXXX 方法抛出 EOFException。而读入批量字节的 read(byte[]) 方法在遇到文件尾时返回 -1，否则返回实际读取的字节数。

10.2.4　字符流

字符按照字符编码规则定义的字节序列，一个字符可能是一字节，也可能是多字节。如果按 UTF-8 编码写入的字符没有按 UTF-8 编码读出，那么就可能会产生"乱码"。纯文本文件（plain text files）、HTML 文档（Hypertext Markup Language documents）和 Java 源文件都按字符流读写。存储字符序列的文件就称为文本文件。

在 JVM 和 .class 文件中都是使用 Unicode 字符集（Unicode character set）上定义的"代码点（code point）"表示字符。当把程序中的字符写入磁盘文件，就需要把代码点转换为计算机系统默认的或参数指定的字符集和字符编码。

不同的操作系统使用不同的字符标识文本文件中一行的结束（称为行尾）。Windows 操作系统则使用回车和换行两个字符（"\r\n"）；UNIX 操作系统，包括 Linux，使用换行符（line feed，'\n'）；macOS 使用回车符（carriage return，'\r'）。

与以字节为单位进行读写的 InputStream 和 OutputStream 相对应,java.io.Reader 和 java.io.Writer 以字符为单位进行读写。Reader 的子类有 InputStreamReader、BufferedReader、CharArrayReader、StringReader 等。FileReader 是 InputStreamReader 的子类,用来按照指定或默认的字符集和字符编码对字节流解码读取文本文件中的字符。如构造方法 FileReader(File)按默认字符集读文件;FileReader(File,Charset)按参数指定的字符集读文件。

java.io.Writer 的子类有 OutputStreamWriter、BufferedWriter、PrintWriter、StringWriter 等。java.io.FileWriter 是 OutputStreamWriter 的子类,用来写文本文件。FileWriter 有 3 个 write 方法:write(int c)、write(char[] cbuf,int off,int len)和 write(String str,int off,int len)。其中 off 表示从文件偏移位置,len 表示字符个数。

可以把文件名作为字符串参数使用 FileWriter(String)构造方法创建对象,也可以先创建 File 对象,以 File 对象作为参数使用 FileWriter(File)构造方法创建对象。如果参数指定的文件已经存在,则原文件被清除。如果希望保留原文件的内容,向其中追加数据,则在调用构造方法时增加一个参数 true 指示进行追加。即使用构造方法 FileWriter(String,boolean)或 FileWriter(File,boolean)。

源文件 10-14 演示了如何使用 FileWriter 向文件 cars.txt 中写入下面两行字符:

京 A12345
冀 AB5678

源文件 10-14　FileWriterDemo.java

```
1 import java.io.FileWriter;
2 import java.io.IOException;
3
4 public class FileWriterDemo {
5     public static void main(String[] args) throws IOException {
6         String file = "cars.txt";
7         FileWriter fos = null;
8         fos = new FileWriter(file);
9         fos.write("京 A12345", 0, "京 A12345".length());
10        fos.write("\r\n");
11        fos.write("冀 AB5678", 0, "冀 AB5678".length());
12        fos.close();
13    }
14 }
```

在源文件 10-14 中第 8 行直接创建了 FileWriter 对象用来访问当前文件夹中的磁盘文件 cars.txt。第 9 行到第 11 行 FileWriter 对象执行 write 方法写入字符。当第 1 行写入完毕,继续让 FileWriter 执行 write 方法写入回车和换行符,以指示当前行的末尾。当使用字符流对象完成文本文件写入后可以使用其他文本编辑器打开该文件,比如 Windows 中的记事本。

FileReader 类 read 方法负责从文本文件中读取字符,其返回值是 0 到 65 535 之间的无符号整数,表示该字符的 Unicode 代码点。如果遇到文件尾,则 read 方法返回-1。

源文件 10-15 使用 FileReader 读取源文件 10-14 写入的文本文件 cars.txt。

源文件 10-15　FileReaderDemo.java

```java
1  import java.io.FileReader;
2  import java.io.IOException;
3
4  public class FileReaderDemo {
5      public static void main(String[] args) throws IOException {
6          String file = "cars.txt";
7          FileReader fis = null;
8          int ch;
9          fis = new FileReader(file);
10         System.out.println(fis.getEncoding());
11         ch = fis.read();
12         while (ch != -1) {
13             System.out.print((char) ch);
14             ch = fis.read();
15         }
16         fis.close();
17     }
18 }
```

源文件 10-15 首先创建了 FileReader 对象用来访问当前文件夹中的磁盘文件 cars.txt。然后使用 while 循环逐个读入字符并输出到屏幕。由于 read 方法返回 0～65 535 的整数，所以需要使用强制类型转换将其转型为 char 类型。如果 read 方法的返回值是－1，即到达文件末尾，则 while 循环结束。程序使用 getEncoding 方法返回文件使用的字符编码名称。程序输出结果是：

```
UTF8
京 A12345
冀 AB5678
```

其中，"UTF8"是运行程序的操作系统默认的字符编码，也是文本文件 cars.txt 使用的字符编码。

在某些情形下，如果需要从字节输入流 InputStream 读取字符数据，或者使用字节输出流 OutputStream 写入字符数据，那么就得先把字节流转换为字符流，或者把字符流转换成字节流。

InputStreamReader 负责把字节流转换成字符流，其构造方法签名如下：

```
InputStreamReader(InputStream)
```

OutputStreamWriter 负责把字节流转换成字符流，其构造方法签名如下：

```
OutputStreamWriter(OutputStream)
```

由于历史的原因，标准输入流对象 System.in 就是字节流 InputStream 类型。为了从标准输入中读取字符，就可以将其转换为字符流：new InputStreamReader(System.in)。

与字节缓冲流 BufferedInputStream 和 BufferedOutputStream 类似，为了提供输入输出效率，字符流也可以进行缓冲。BufferedReader 和 BufferedWriter 类就提供了内部字符缓冲区。BufferedReader 类读取字符的方法有：read 方法从字符流中读取 1 个字符，当遇到文件末尾时返回－1；read(char[]cbuff,intoff,int len)方法读取 len 个字符，并存储到从索引位置 off 开始的字符数组 cbuff 中,当遇到文件末尾时返回－1。readLine 方法从字符流中读取 1 行并以字符串对象返回,但是该行中不含行分隔符,如果到达文件尾则返回 null。BufferedWriter 类写入字符的方法有:write(int)写入 1 个字符;write(char[]cbuff, intoff, int len)方法把一个字符数组 cbuff 中字符从指定索引位置 off 开始,写入 len 个字符;write(String s, int off, int len)方法从指定索引位置 off 开始,写入字符串 s 中的 len 个字符;newLine()方法写入行分隔;write(String str)等价于 write(str, 0, str.length())。

源文件 10-16 使用 BufferedReader 的 readLine 方法逐行读取源文件 10-14 建立的文本文件 cars.txt,并输出到标准输出(System.out)。

源文件 10-16　BufferedReaderDemo.java

```
1 import java.io.FileReader;
2 import java.io.BufferedReader;
3 import java.io.IOException;
4
5 public class BufferedReaderDemo {
6     public static void main(String[] args) throws IOException {
7         String file = "cars.txt";
8         BufferedReader br = null;
9         String aLine;
10
11        br = new BufferedReader(new FileReader(file));
12        while ((aLine = br.readLine()) != null) {
13            System.out.println(aLine);
14        }
15        br.close();
16    }
17 }
```

注意:先创建 FileReader 对象用来访问磁盘文件 cars.txt,然后创建 BufferedReader 对象以缓冲方式访问 FileReader 对象。BufferedWriter 类中的 newLine 方法目的是提供一个与平台无关的方式写入行分隔符。前文提到,不同平台的文本文件使用的行分隔符不同:UNIX 使用回车符,macOS 使用换行符,Windows 使用回车换行两个字符。newLine 方法能够识别程序运行系统的行分隔符参数,在不同的系统上写入不同的行分隔符。

10.3　Scanner 类和 PrintWriter 类

当需要从文本文件中读取字符串或各种类型的数值时,使用 java.util.Scanner;当需要把字符串或各种类型的数值写入文本文件时,使用 java.io.PrintWriter。

为了简化对 8 种基本数据类型数据的读取,java.util.Scanner 提供了一系列方法从字符流中解析并读取指定数据类型的数据。Scanner 的常用构造方法有:

- Scanner(File source)
- Scanner(File source, String charsetName)
- Scanner(InputStream source)
- Scanner(InputStream source, String charsetName)
- Scanner(String source)

可见，Scanner 对象以字节输入流 InputStream 对象、File 对象，或者 String 对象作为数据源。

源文件 10-17 演示了如何使用 Scanner 对象逐行读取文本文件。

源文件 10-17　ScannerExample.java

```
1  import java.io.IOException;
2  import java.io.File;
3  import java.util.Scanner;
4
5  public class ScannerExample {
6      public static void main(String[] args) throws IOException {
7          File fromFile = new File("cars.txt");
8          Scanner reader = null;
9          reader = new Scanner(fromFile);
10         //逐行进行处理
11         while (reader.hasNextLine()) {
12             System.out.println(reader.nextLine());
13         }
14         reader.close();
15     }
16 }
```

在源文件 10-17 中的第 9 行创建了 Scanner 对象访问源文件 10-14 磁盘文件 cars.txt。第 11 行通过 Scanner 对象的 hasNextLine 方法判断是否还有下一行。如果输入流中还有一行，第 12 行首先使用 nextLine 方法将其以字符串形式返回。返回的字符串立即通过 System.out 对象的 println 方法输出显示。这个过程直到 hasNextLine 方法返回 false。

Scanner 对象的主要作用在于从字符流中解析各种数据类型的数据。各个数据项之间使用分隔符隔开。默认的分隔符是空白符。分隔符把字符流分隔成不同词元(token)。这样，字符流就可以看作是词元流。比如以制表符分隔的字符序列：

```
-1    16    3.14    true    ABC
```

就被 Scanner 对象识别为词元序列：

```
"-1","16","3.14","true"和"Donald"
```

使用 nextXXX 方法把词元序列中的词元解析为不同数据类型的数据。部分 nextXXX 方法的功能见表 10-2。

表 10-2　nextXXX 方法

方法名	功能
next	把下个词元以字符串对象返回
nextByte	把下个词元解析为 byte 类型数据并返回
nextShort	把下个词元解析为 short 类型数据并返回
nextInt	把下个词元解析为 int 类型数据并返回
nextLong	把下个词元解析为 long 类型数据并返回
nextBoolean	把下个词元解析为 boolean 类型数据并返回
nextFloat	把下个词元解析为 float 类型数据并返回
nextDouble	把下个词元解析为 double 类型数据并返回

通常在使用 nextXXX 方法之前先使用 hasNextXXX 方法判断是否是想要的数据。部分 hasNextXXX 方法如表 10-3 所示。

表 10-3　hasNextXXX 方法

方　法　名	功　能
hasNext	如果还有词元返回 true
hasNextByte	如果还有词元，而且该词元能够被解析为 byte 类型数据则返回 true
hasNextShort	如果还有词元，而且该词元能够被解析为 short 类型数据则返回 true
hasNextInt	如果还有词元，而且该词元能够被解析为 int 类型数据则返回 true
hasNextLong	如果还有词元，而且该词元能够被解析为 long 类型数据则返回 true
hasNextBoolean	如果还有词元，而且该词元能够被解析为 boolean 类型数据则返回 true
hasNextFloat	如果还有词元，而且该词元能够被解析为 float 类型数据则返回 true
hasNextDouble	如果还有词元，而且该词元能够被解析为 double 类型数据则返回 true

Scanner 不仅支持读取基本数据类型的数据，还支持 BigInteger、BigDecimal 等类型的数据。

假设文本文件 in.txt 中存放了以制表符隔开的数字字符序列如下：

```
-1    16    3.14    true    ABC
```

源文件 10-18 演示了如何使用 Scanner 对象从 in.txt 文件读取不同类型的数据。

源文件 10-18　ScannerDemo.java

```
1 import java.io.File;
2 import java.io.IOException;
3 import java.util.Scanner;
4
5 public class ScannerDemo {
6     public static void main(String[] args) throws IOException {
7         String filePath = "in.txt";
8         Scanner sc = null;
9         sc = new Scanner(new File(filePath));
10        System.out.println(sc.nextByte());
11        System.out.println(sc.nextInt());
12        System.out.println(sc.nextDouble());
13        System.out.println(sc.nextBoolean());
14        System.out.println(sc.next());
15        sc.close();
16    }
17 }
```

在源文件 10-18 中第 9 行创建了 Scanner 对象来访问磁盘文本文件 in.txt。然后使用不同的 nextXXX 方法逐个读取数据项。程序的输出如下：

```
-1
16
3.14
true
ABC
```

当数据源中有大量数据项时,通常使用循环结构来读取数据:

```
Scanner sc = new Scanner(…);
while(sc.hasNext()) {
    sc.nextXXX();
    //...
}
```

即首先创建 Scanner 对象,然后使用 while 循环完成数据读取。使用 hasNext 方法控制循环;在循环体中使用 nextXXX 方法完成数据项的读取。

假如在文本文件 data.txt 有空格分隔的数据项:80 78 89 95 100 90,那么可使用循环结构读取,如源文件 10-19 所示。

源文件 10-19　ScannerLoopDemo.java

```
1  import java.io.File;
2  import java.io.IOException;
3  import java.util.Scanner;
4
5  public class ScannerLoopDemo {
6      public static void main(String[] args) throws IOException {
7          String filePath = "data.txt";
8          Scanner sc = null;
9          sc = new Scanner(new File(filePath));
10         while (sc.hasNextInt()) {
11             System.out.println(sc.nextInt());
12         }
13         sc.close();
14     }
15 }
```

在 Java 中,System 类中定义的静态变量 System.in 引用字节输入流 InputStream 类型的对象,这个对象封装了来自标准输入(键盘)的字节流。Scanner 对象可把来自标准输入的字节流转换成字符流,再变换成记号流,从而完成各种类型的数据的读取。

从表 10-2 可以看到,Scanner 类中没有提供方法读取 char 类型数据,即没有 nextChar 方法。

为了读取单个字符,一种办法是先按照字符串读入一个词元,然后使用字符串对象的 chartAt 方法提取单个字符,如源文件 10-20 所示。

源文件 10-20　CharTest.java

```
1  import java.util.Scanner;
2
3  public class CharTest {
4      public static void main(String[] args) {
5          Scanner in = new Scanner(System.in);
6          //从字符串中截取首个字符
7          System.out.println("Enter a character");
8          String token = in.next();
9          char ch = token.charAt(0);
10         System.out.println("The character was: " + ch);
11     }
12 }
```

运行程序,按照提示输入"Y",交互显示结果为:

```
Enter a character
Y
The character was: Y
```

面向字符输出的 PrintWriter 重载了 print、printf 等方法,用以输出各种类型的数据。源文件 10-21 中的程序向文本文件 out.txt 中写入了一个字符串"Hi,",一个整数 16,一个浮点数 3.14 和一个布尔值。

源文件 10-21　PrintWriterExample.java

```java
1  import java.io.FileWriter;
2  import java.io.IOException;
3  import java.io.PrintWriter;
4  
5  public class PrintWriterExample {
6      public static void main(String[] args) throws IOException {
7          PrintWriter pw = new PrintWriter(new FileWriter("out.txt"));
8          pw.print("Hi,");
9          pw.print(16);
10         pw.print(3.14);
11         pw.println(true);
12         pw.close();
13     }
14 }
```

使用文本编辑器打开文本文件 out.txt,看到里面有两行:

```
1  Hi,163.14true
2
```

其中第 2 行是因为源文件 10-21 的第 11 行使用了 println 方法输出而不是 print 方法。println 方法名中的 ln 读作 line,意思是输出完数据后换行。

也可以使用 PrintWriter 的 printf 方法按照 C 语言的习惯输出,如源文件 10-22 所示。

源文件 10-22　PrintWriterPrintfDemo.java

```java
1  import java.io.FileWriter;
2  import java.io.IOException;
3  import java.io.PrintWriter;
4  
5  public class PrintWriterPrintfDemo {
6      public static void main(String[] args) throws IOException {
7          PrintWriter pw = new PrintWriter(new FileWriter("out2.txt"));
8          pw.printf("%s%d%5.2f%B", "Hi,", 16, 3.14, true);
9          pw.close();
10     }
11 }
```

在文本编辑器中打开 out2.txt,可以看到输出结果:

```
1 Hi,16 3.14TRUE
```

printf 方法的头部声明如下:

```
printf(String format, Object… args)
```

format 称为格式串,后面是 0 个或若干个参数。

格式串由可选的固定文本(fixed text)以及 1 个或多个格式说明符(format specifiers)组成。格式说明符从百分号开始(%)。%后面是转换码(conversion code),指示按何种数据类型输出相应的参数。例如"%f"中的"f"即是转换码,用以指示按浮点数输出。

在百分号和转换码之间可以指定数据的宽度和小数点后数字的位数。例如,格式说明符"%9.5f"的意思是:以小数点前 3 位,小数点后 5 位的格式输出,如果含小数点不足 9 位字符,则右对齐,左补空格。"%-9.5f"以小数点前 3 位,小数点后 5 位的格式输出,如果含小数点不足 9 位字符,则左对齐。

以语句 System.out.printf("The value of PI is:%9.5f%n", Math.PI);为例,printf 方法中参数的说明如图 10-6 所示。其中,"The value of PI is:%9.5f%n"称为格式串。格式串中的"The value of PI is:"称为固定文本。格式串中的"%9.5f"称为格式描述符,用以指示按照此格式输出相应的参数 Math.PI。"%n"也是格式描述符,指示输出平台特定的行分隔符。

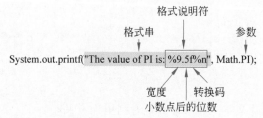

图 10-6 printf 方法的参数

除了 f 外,其他的转换码如表 10-4 所示。

表 10-4 转换码

转换码	含 义	转换码	含 义
c,C	Unicode 字符	s,S	字符串
d	十进制整数	x,X	十六进制整数
e,E	科学记数法,如 1.23e-03 或 1.23E-03	%	格式串中的"%%"输出单个"%"('\u0025')
b,B	逻辑值,true 或 false	n	换行符
o	八进制整数	e,E	科学记数法表示的浮点数

格式描述符的完整形式如下:

%[参数索引$][标志][宽度][.小数点后位数]转换码

其中,"参数索引"指示当前格式描述符应用于第几个参数。比如%2$指示当前的格式描述符应用于第 2 个参数。可以使用也可以不使用"参数索引"。

"标志"指示对齐方式和填充方式。例如,"%+f"中的+表示包含正负符号,"%-f"中的-表示左对齐,"%05d"表示使用 0 填充而不是使用默认的空格填充。

源文件 10-23 演示了各种格式说明符的效果。

源文件 10-23　PrintfTest.java

```java
import java.util.Locale;
public class PrintfTest {
    public static void main(String[] args) {
        long n = 123456;
        System.out.printf("%d%n", n);                              //123456
        System.out.printf("%08d%n", n);                            //00123456
        System.out.printf("%+8d%n", n);                            //□+123456
        System.out.printf("%,8d%n", n);                            //□123,456
        System.out.printf("%+,8d%n%n", n);                         //+123,456

        double pi = Math.PI;
        System.out.printf("%f%n", pi);                             //3.141593
        System.out.printf("%.3f%n", pi);                           //3.142
        System.out.printf("%10.3f%n", pi);                         //□□□□□3.142
        System.out.printf("%-10.3f%n", pi);                        //3.142□□□□□
        System.out.printf(Locale.FRANCE, "%-10.4f%n%n", pi);       //3,1416□□□□

        double x = 123.123456789;
        System.out.printf("%e%n", x);                              //1.231235e+02
        System.out.printf("%10.3e%n", x);                          //1.231e+02

        System.out.printf("%12s%n","ABCD1234");                    //□□□□ABCD1234
        System.out.printf("%-12s%n","ABCD1234");                   //ABCD12348□□□□
    }
}
```

为了方便阅读，源文件 10-23 把 printf 方法的输出以注释方式写在语句的右侧。其中"□"表示空格符。

10.4　对象串行化

把内存中的对象写入文件、网络等输出流的过程称为对象串行化（serialization）。对象串行化把对象变换成字节流。从文件、网络等输入流读入并在内存中恢复对象的过程称为对象反串行化（deserialization）。只有实现了 Serializable 接口的对象才能被串行化。ObjectOutputStream 和 ObjectInputStream 负责对象串行化和反串行化。ObjectOutputStream 是 Outputstream 的子类；ObjectInputStream 是 InputStream 的子类。ObjectOutputStream 的 writeObject 方法把对象的实例变量的值写入输出流；ObjectInputStream 的 readObject 方法从输入流中重建对象。源文件 10-24 演示了如何进行对象串行化和反串行化。

源文件 10-24　SerializationDemo.java

```java
1 import java.io.File;
2 import java.io.FileInputStream;
3 import java.io.FileOutputStream;
4 import java.io.IOException;
5 import java.io.ObjectInputStream;
6 import java.io.ObjectOutputStream;
```

```java
 7 import java.io.Serializable;
 8
 9 public class SerializationDemo {
10     public static void main(String[] args) {
11         Car stObj = new Car("A666", 18);
12         File objFile = new File("cars.ser");
13
14         try (FileOutputStream fos = new FileOutputStream(objFile);
15              ObjectOutputStream oos = new ObjectOutputStream(fos)){
16             oos.writeObject(stObj);
17         }catch(IOException e){
18             e.printStackTrace();
19             return;
20         }
21
22         try (FileInputStream fis = new FileInputStream(objFile);
23              ObjectInputStream ois = new ObjectInputStream(fis)) {
24             stObj = (Car) ois.readObject();
25         }catch(IOException | ClassNotFoundException e){
26             e.printStackTrace();
27             return;
28         }
29         System.out.println("The Car comes back: " + stObj);
30     }
31 }
32
33 class Car implements Serializable {
34     private String licenceNumber;
35     private int weight;
36
37     Car(String licenceNumber, int weight) {
38         this.licenceNumber =   licenceNumber;
39         this.weight = weight;
40     }
41     @Override
42     public String toString() {
43         return "Car [licenceNumber=" + licenceNumber
44             + ", weight=" + weight  + "]";
45     }
46
47 }
```

源文件 10-24 中首先在第 33 行声明了汽车类 Car。这个包私有的类有两个私有成员变量、一个包私有的构造方法和一个覆盖父类的方法 toString。在第 11 行首先创建了一个 Car 类型的对象，这个对象有两个属性：licenceNumber 和 weight，值分别是字符串对象"A666"和整数 18。程序试图把 Car 对象写入文件 cars.ser；写入成功后再从文件 cars.ser 读出该对象。

第 14 行使用 try-with-resources 结构访问当前文件夹磁盘文件 cars.ser。让 ObjectOutputStream 对象执行 writeObject 方法把 Car 对象写入输出流。

接下来程序进行对象反串行化。第 22 行在 try-with-resources 结构中让 ObjectInputStream 对象执行 readObject 方法把 Car 对象从输入流读入。最后输出读入的对象的字符串表示。

程序输出结果为:

```
The Car comes back: Car[licenceNumber=A666, weight=18]
```

注意:

(1) 由于 readObject 方法返回值类型为 Object,所以需要将其转换为 Car 类型。
(2) Car 类必须实现 Serializable 接口。
(3) 如果对象的成员变量是引用类型,Java 自动确保该变量引用的对象只有一份副本被串行化。
(4) 第 25 行使用"|"分隔了多个异常类型,减少了 catch 子句。

10.5 字符集和 Unicode

美国信息交换标准编码(American standard code for information interchange,ASCII)是一个 7 位的编码方案,共 128 个字符。显然,使用 ASCII 无法处理汉字。Unicode 试图为在计算机乃至网络上处理各种语言中的字符提供解决方案。最新的 Unicode 版本(V6.2,2012)定义了 110 182 个来自不同语言的字符,包括英语、汉语、俄语、阿拉伯语、希腊语等。Unicode 字符集把每个字符映射到一个整数,称为代码点(code point)。代码点从 0x000000 开始到 0x10FFFF 结束共 1 114 112 个。通常使用"U+XXXX"来引用代码点。其中 XXXX 是十六进制数字。例如 U+0041 就是字符 A 的代码点。

Unicode 把代码点分割到 17 个平面上,每个平面有 2^{16}(65 536)个代码点。每个平面中的代码点都是 xx0000 至 xxFFFF,其中 xx 是从 00 到 10 的十六进制值。平面 0(U+0000~U+FFFF)称为基本多语言平面(Basic Multilingual Plane,BMP),其中包含了使用频率最高的字符。

Unicode 是一个字符集而不是字符编码(encoding)。也就是说,虽然 Unicode 定义了字母 A 的代码点是 U+0041,但并没有定义使用几个字节存储,高位是否放在高地址等。UTF-8 是一个 8bit 变长的编码模式,是 Unicode 文本事实上的编码标准。UTF-8 最大特点是与 ASCII 兼容。

表 10-5 UTF-8 编码模式

位数	最后代码点	字节 1	字节 2	字节 3	字节 4	字节 5	字节 6
7	U+007F	0xxxxxxx					
11	U+07FF	110xxxxx	10xxxxxx				
16	U+FFFF	1110xxxx	10xxxxxx	10xxxxxx			
21	U+1FFFFF	11110xxx	10xxxxxx	10xxxxxx	10xxxxxx		
26	U+3FFFFFF	111110xx	10xxxxxx	10xxxxxx	10xxxxxx	10xxxxxx	
31	U+7FFFFFFF	1111110x	10xxxxxx	10xxxxxx	10xxxxxx	10xxxxxx	10xxxxxx

单字节编码用于编码 ASCII 字符集,其值从 0 到 127。即对于 ASCII 字符,UTF-8 编码与 ASCII 编码相同。这些字节的最高位总是 0。

大于127的代码点使用多字节编码。多字节编码由1个前导字节（leading byte）和若干个后随字节组成。前导字节最高2位，或最高3位，或最高4位，或最高5位，或最高6位为1，后面紧跟1个0。前导字节中0之前1的个数指示了多字节编码的字节个数，有几个1就有几字节。所有后随字节的最高两位都是"10"。

操作系统的区域属性和默认字符集决定了Java虚拟机的默认字符集。例如，汉字"您"的Unicode代码点是"60A8"，对应的UTF-8编码是"E682A8"；对应的GBK编码是"C4 FA"。字符串"ABC您好"按照UTF-8编码的结果是41 42 43 E6 82 A8 E5 A5 BD；而按照GBK编码的结果是41 42 43 C4 FA BA C3。注意，对于字母ABC，UTF-8的编码与ASCII编码相同；而对于汉字，UTF-8则使用三字节编码而GBK使用双字节编码。

源文件10-25演示了使用UTF-8编码从文本文件input.txt中读取一行，然后写入文本文件output.txt中。

源文件10-25　CharsetTest.java

```java
1  import java.io.File;
2  import java.io.PrintWriter;
3  import java.io.IOException;
4  import java.util.Scanner;
5
6  public class CharsetTest {
7      public static void main(String[] args) {
8          try (Scanner in = new Scanner(new File("input.txt"), "UTF-8");
9               PrintWriter out = new PrintWriter(new File("output.txt"),
                     "UTF-8")) {
10             out.println(in.nextLine());
11         } catch (IOException e) {
12             e.printStackTrace();
13         }
14     }
15 }
```

10.6　记录

如果某对象在创建后状态一直保持不变，则称为"不变对象"。保留字final禁止修改变量的值：

```
final String licenceNumber = "1234";
name = "bael...";
```

但是，final只是禁止更改变量的值，并不能禁止更改它所引用的对象的状态：

```
final List<String> strings = new ArrayList<>();
strings.add("1234");
```

如果一个磁盘文件由若干行组成，每行由固定顺序的若干个值组成，那么文件中的每一行称为一条记录（record），记录中有若干字段（field），值就是某字段的值。许多情况下记录

是不变的。为了实现不变性,可使用以下方法设计为记录类 Record:

每个字段由 private 和 final 修饰;

每个字段仅有 getter 方法,无 setter 方法;

全参数公共构造方法;

equals 方法,对同一类的对象若所有字段匹配则返回 true;

hashCode 方法,当所有字段匹配时返回相同的值;

toString 方法,包括类的名称、每个字段的名称及其对应的值。

例如,源文件 10-26 就是一个实现了不变性的汽车类 Car,每辆汽车有 4 个属性:车牌号码、外廓尺寸(长、宽和高)。

源文件 10-26 Car.java

```
1  import java.util.Objects;
2
3  public class Car {
4
5      private final String licenceNumber;
6      private final int length, width, height;
7
8      public Car(String licenceNumber, int length, int width, int height) {
9          this.licenceNumber = licenceNumber;
10         this.length = length;
11         this.width = width;
12         this.height = height;
13     }
14
15     @Override
16     public int hashCode() {
17         return Objects.hash(licenceNumber, length, width, height);
18     }
19
20     @Override
21     public boolean equals(Object obj) {
22         if (this == obj) {
23             return true;
24         } else if (!(obj instanceof Car)) {
25             return false;
26         } else {
27             Car other = (Car) obj;
28             return Objects.equals(licenceNumber, other.getLicenceNumber())
29                 && Objects.equals(length, other.getLength())
30                 && Objects.equals(width, other.getWidth())
31                 && Objects.equals(height, other.getHeight());
32         }
33     }
34
35     @Override
36     public String toString() {
37         return "Car [Licence Number = " + licenceNumber
```

```
38              + ", Length=" + length
39              + ", Width=" + width
40              + ", Height=" + height + "]";
41      }
42
43      //标准的 getters
44      public String getLicenceNumber() {
45          return licenceNumber;
46      }
47
48      public int getLength() {
49          return length;
50      }
51
52      public int getWidth() {
53          return length;
54      }
55
56      public int getHeight() {
57          return length;
58      }
59  }
```

使用 java.lang.Record 重构源文件 10-26 能够获得如下益处。

(1) 简化代码。Record 可以自动实现 equals()、hashCode() 和 toString() 方法,大大减少了开发者手动编写这些方法的工作量。

(2) 提高可读性。Record 直接通过字段列表来声明其数据结构,简单明了。

(3) 保证不变性。Record 中的所有字段默认都是 final 的,因此创建后不能修改。

(4) 契约清晰。Record 明确表明其目的是存储一组相关联的值,并且没有行为(除了构造方法、getter 方法和必要的 equals/hashCode/toString 方法)。

对源文件 10-26 的重构如源文件 10-27 所示。

源文件 10-27　Car.java

```
public record Car(String licenceNumber, int length, int width, int height) {

}
```

源文件 10-28 演示了对源文件 10-27 中声明的记录 Car 的使用。在第 8 行和第 9 行分别创建了两条记录并通过 c 和 d 来引用;第 11 行调用了记录中的 getter 方法 getLicenceNumber();第 12 行调用了 equals 方法;第 13 行调用了 toString() 方法。这些方法均由编译器生成。

源文件 10-28　TestCar.java

```
1 public class TestCar {
2     public static void main(String[] args) {
3         String licenceNumber = "1234";
4         int length = 1456;
5         int width = 890;
```

```
 6          int height = 567;
 7
 8          Car c = new Car(licenceNumber, length, width, length);
 9          Car d = new Car(licenceNumber, length, width, length);
10
11          System.out.println(c.licenceNumber());
12          System.out.println(c.equals(d));
13          System.out.println(c);
14      }
15 }
```

在源文件10-26第28行中使用了java.util.Objects,该类自JDK 1.7版本开始提供,为操作对象提供了静态实用程序方法,如判断两个对象是否相等的 equals(Object a,Object b)、判断是否为空的 isNull(Object obj)、把对象转换为字符串的 toString(Object o)等。

第 10 章　章节测验

第11章 多线程程序设计

小明的班里有30名同学。在数学课上,老师先让一名同学在黑板上演算一道数学题。然后老师问同学们:"谁能说一说黑板上的证明过程有没有问题?"然后同学们纷纷举手,表示要发言。小明也举了手。老师说:"小明说说看。"小明站起来,大声说:"在证明三角形全等时,条件没有写全"。老师说:"说得对。"在这个场景中,30名同学以及1位老师共同完成教学活动。而教室环境下教学活动的特点是:每个时刻只允许一个人讲话。所以,当需要同学们发言时,必须先举手示意,经过老师同意再站起来发言。

多个程序可以在计算机中一起运行,就好比多个学生在教室一起上课。正如一个课堂中只能有一个人发言,在单核计算机中每时刻也只有一个程序运行。

11.1 进程和线程

程序在计算机上的运行实例称为进程(process)。比如在 Windows "同时"运行两个 Notepad3 应用程序,如图11-1所示。这两个看起来在"同时"运行的 Notepad3 应用就是两个进程,在 Windows 操作系统中称为"任务(task)"。

虽然安装在磁盘上的 Notepad3 程序只有一份,但可以运行多个 Notepad3 实例,即多个 Notepad3 进程。当关闭计算机,Notepad3 的程序仍然在磁盘上,下次还可以启动;但 Notepad3 进程全部消失。

操作系统为每个进程都指派了一个 ID,称为进程号。每个进程都有自己的内存空间和运行状态,称为进程的上下文(context)。进程中还可以创建多个线程(thread)来协作完成一个任务。

线程就是进程中可被调度运行的指令序列。进程中的线程共享进程的内存空间。进程间的切换由操作系统管理,需要花费一定的时间和空间代价完成;而线程间的切换代价要小得多。如何在应用进程中启动和管理多个线程以高效地完成一定的任务,就是多线程程序

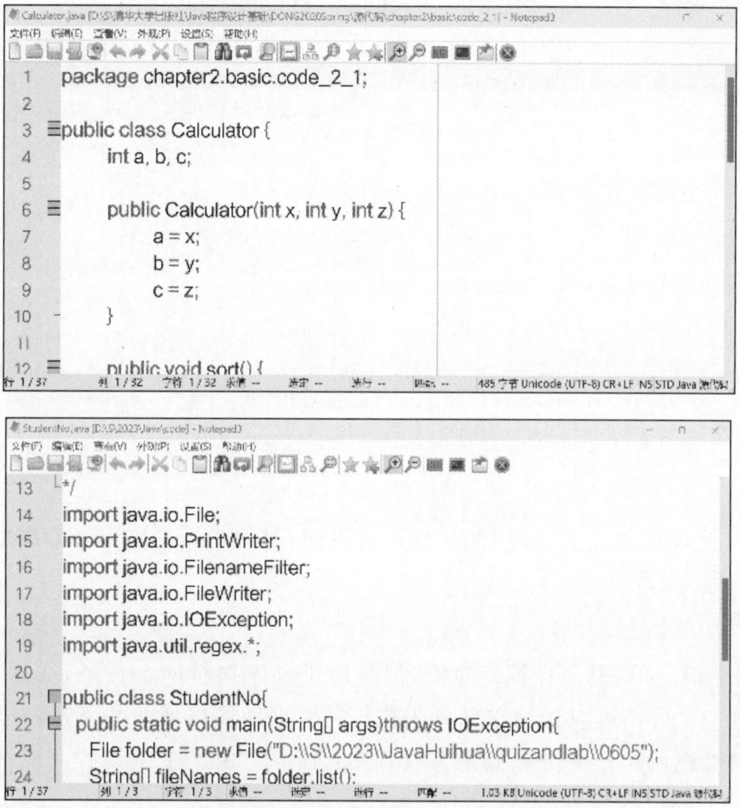

图 11-1 两个 Notepad3 进程

设计(multithreaded programming)。多线程是 Java 平台本质的特征之一。

每个应用进程中至少有一个线程,称为主(main)线程。线程中可以创建新的线程。创建者称为父线程,被创建的线程称为子线程。例如,主线程启动后,可以创建并启动线程 A。这样,线程 A 就与主线程一起处于可运行状态。"可运行"状态指只要允许就可以在中央处理器上执行的状态。好比是同学们都在举手,只要老师允许,就可以发言一样。线程 A 也可以创建和启动线程 B,这样,进程中就有三个线程处于可运行状态。多线程程序设计更好地发挥了多核中央处理器的性能。注意,在多线程情形中,无法预知某个时刻哪个线程在中央处理器上执行。

下面使用一个例子来说明多线程程序设计的必要性。这个例子要求设计程序完成两个任务。任务 1 是计算 Fibonacci 序列中第 50 个数是什么;任务 2 是计算 −50 的绝对值是多少。Fibonacci 序列是一个整数序列:0,1,1,2,3,5,8,13,……。在 Fibonacci 序列中,前两个整数是 0 和 1,后面的整数是其前两个整数之和。

根据 Fibonacci 序列的数学定义,计算序列中第 50 个数需要知道第 49 个数和第 48 个数,然后相加即可。要想知道第 49 个数,就需要知道第 48 个数和第 47 个数,……。数越大,这个过程越耗时。计算耗时的任务也称为"计算敏感"的任务。源文件 11-1 在 FibonacciNumber 类里设计了计算序列中第 n 个数的类方法 fib。在 main 方法中依次完成两个任务:计算序列中第 50 个数、计算 −50 的绝对值。

源文件 11-1　FibonacciNumber.java

```java
public class FibonacciNumber {
    private static long fib(long n) {
        if (n == 0) {
            return 0L;
        }
        if (n == 1) {
            return 1L;
        }
        return fib(n - 1) + fib(n - 2);
    }

    public static void main(String[] args) throws InterruptedException {
        System.out.println("任务1: " + fib(50));
        System.out.println("任务2: " + Math.abs(-50));
    }
}
```

运行程序,发现程序过了一段时间(计算机不同,时间长短不同)后,输出如下结果:

任务1: 12586269025
任务2: 50

虽然任务2很简单,只是计算绝对值,但是由于顺序执行两个任务,必须等任务1完成后,才能做任务2。可不可以让简单任务先执行完？即使简单任务并不是先提交运行。

源文件11-2把两个任务分派到两个线程中：任务1由一个子线程完成；任务2由另一个线程完成。

源文件 11-2　FibonacciTask.java

```java
1  public class FibonacciTask {
2      private static long fib(long n) {
3          if (n == 0) {
4              return 0L;
5          }
6          if (n == 1) {
7              return 1L;
8          }
9          return fib(n - 1) + fib(n - 2);
10     }
11
12     public static void main(String[] args) throws InterruptedException {
13         System.out.println("主线程开始执行...");
14
15         Thread threadFib = new Thread(() -> {
16             System.out.println("子线程执行任务1: " +  fib(40));
17         });
18         Thread threadAbs = new Thread(() -> {
19             System.out.println("子线程执行任务2: " + Math.abs(-50));
20         });
21         threadFib.start();
22         threadAbs.start();
23         System.out.println("主线程终止");
24     }
25 }
```

运行程序,会看到如下结果:

```
主线程开始执行...
主线程终止
子线程执行任务 2: 50
子线程执行任务 1: 102334155
```

在第 15 行和第 18 行主线程首先创建两个子线程,分别完成计算 Fibonacci 序列第 40 个数的任务和计算－50 绝对值的任务。在第 21 行和第 22 行启动子线程开始运行;在子线程运行期间,主线程也在运行,但很快终止。接着计算绝对值的任务 2 很快完成,输出结果 50;等待一段时间后任务 1 也完成了。这样,即使简单任务并不是先提交运行也可能先运行结束。

下面就介绍如何创建线程,如何启动线程,如何管理线程。

11.2 创建线程

创建线程有三种方法:继承 java.lang.Thread 类并覆盖 run 方法;实现 java.lang.Runnable 接口并覆盖 run 方法;或者实现 java.util.concurrent.Callable 接口。本节先介绍前两种方法。

Thread 类也是 Runnable 接口的实现类,其关系如图 11-2 所示。Thread 类中还提供了让线程启动 start()、中断处于休眠状态的线程 interrupt()、查询线程的名字 getName()、设置线程的名字 setName()、判断线程是否已经终止运行 isAlive()、设置线程的优先级 setPriority()、等待线程结束的 join() 等实例方法。还有查

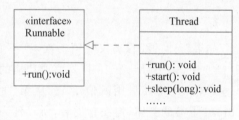

图 11-2 Thread 类与 Runnable 接口

询正在执行的线程静态方法 currentThread()、休眠 sleep()、主动放弃在中央处理器上执行 yield() 等。自 JDK 1.4 后,不再使用 Thread 类中的 stop()、suspend()、resume() 等方法。

通过继承 Thread 类,重写 run() 方法,把功能安排在 run 方法中来定义线程。例如在源文件 11-3 的第 4 行把计算第 7 个 Fibonacci 数的私有方法 fib 安排在线程的 run 方法中,从而定义了一个新的线程 Fibonacci。Fibonacci 线程通过继承 Thread 类获取了线程所共有的属性和方法。

源文件 11-3 Fibonacci.java

```
1 public class Fibonacci extends Thread {
2     @Override
3     public void run() {
4         long result = fib(7);
5     }
6
7     private long fib(long n) {
8         if (n == 0) {
9             return 0L;
```

```
10      }
11      if (n == 1) {
12          return 1L;
13      }
14      return fib(n - 1) + fib(n - 2);
15  }
16 }
```

在线程Fibonacci中并没有直接把计算Fibonacci序列的代码片段安排在run()方法中。而是首先把计算Fibonacci序列的代码段定义为私有方法,然后在run()方法中调用这个私有方法。这样的设计使程序的可读性高一些。

设计好线程类后就创建线程并运行。在源文件11-4的第3行,通过Fibonacci线程类的无参构造方法创建了线程并使用引用变量t引用这个线程对象。线程被创建后并不会自动运行。接下来程序让线程t执行start()方法进入可运行(RUNNABLE)状态。这样线程t就和当前正处于可运行状态的其他线程一起等待调度和执行。线程的实例方法start()把线程对象安排好之后就立即返回,并不等待线程执行结束。

源文件11-4　TestFibonacci.java

```
1 public class TestFibonacci {
2     public static void main(String[] args) {
3         Thread t = new Fibonacci();
4         t.start();
5     }
6 }
```

如果新建的线程类仅使用一次,则不必命名,直接使用匿名类创建线程对象。在源文件11-5中,第2行到第10行是完成功能的私有方法,第12行到第18行使用内部类创建了线程对象,并将其引用赋值给t。从第13行末的左花括号到第20行的右花括号是一个Thread子类的类体,但没有类名,所以称为匿名类。第19行启动匿名类创建的线程对象。

源文件11-5　TestFibonacciAnonymous.java

```
1 public class TestFibonacciAnonymous {
2     static private long fib(long n) {
3         if (n == 0) {
4             return 0L;
5         }
6         if (n == 1) {
7             return 1L;
8         }
9         return fib(n - 1) + fib(n - 2);
10     }
11
12     public static void main(String[] args) {
13         Thread t = new Thread() {
14             @Override
15             public void run() {
16                 long result = fib(7);
17             }
18         };
19         t.start();
20     }
21 }
```

Thread 类的构造方法中有线程组（ThreadGroup），Runnable 对象和线程的名字（String）等参数，常见的构造方法有 Thread()、Thread(Runnable task)、Thread(String name)、Thread(Runnable task，String name)等。

通过实现 Runnable 接口创建线程有利于程序的扩展性。一般步骤如下。

(1) 定义 Runnable 接口的实现类，实现抽象方法 run()。
(2) 创建实现类的实例，这个实例称为 Runnable 对象。
(3) 以 Runnable 对象为构造方法参数，创建 Thread 对象。
(4) 让 Thread 对象执行 start 方法进入可运行状态。

源文件 11-6　TestFibonacciRunnable.java

```
1 class FibonacciRunnable implements Runnable {
2     public void run() {
3         long result = fib(7);
4     }
5
6     private long fib(long n) {
7         if (n == 0) {
8             return 0L;
9         }
10        if (n == 1) {
11            return 1L;
12        }
13        return fib(n - 1) + fib(n - 2);
14    }
15 }
16
17 public class TestFibonacciRunnable {
18     public static void main(String[] args) {
19         Runnable r = new FibonacciRunnable();
20         Thread t = new Thread(r);
21         t.start();
22     }
23 }
```

在源文件 11-5 中第 1 行通过实现 Runnable 接口定义 FibonacciRunnable 线程。但是 Runnable 对象还不是线程。因此接下来在第 19 行创建 FibonacciRunnable 对象，把 FibonacciRunnable 对象作为构造方法参数创建 Thread 对象。最后让 Thread 对象执行 start 方法启动线程。

在源文件 11-7 中第 3 行到第 8 行使用匿名内部类创建 Runnable 对象并使用该对象作为构造方法参数创建 Thread 对象。

源文件 11-7　TestAnonymousRunnable.java

```
1 public class TestAnonymousRunnable {
2     public static void main(String[] args) {
3         Thread t = new Thread(new Runnable() {
4             @Override
5             public void run() {
6                 long result = fib(7);
7                 System.out.println(result);
```

```
8       }
9
10      private long fib(long n) {
11          if (n == 0) {
12              return 0L;
13          }
14          if (n == 1) {
15              return 1L;
16          }
17          return fib(n - 1) + fib(n - 2);
18      }
19  });
20
21  t.start();
22  }
23
24 }
```

Runnable 对象并不是线程对象,但可以以 Runnable 对象为参数创建线程对象。从第3行末的左花括号到第19行的右花括号是 Runnable 实现类的类体,但该实现类没有类名,是一个匿名类。

在源文件11-8 的第3行到第7行使用 Lambda 表达式创建线程。Lambda 表达式可以用来简洁地定义一个匿名的、实现了 Runnable 接口的类实例。传递一个 Lambda 表达式给 Thread 构造方法时,实际上就是传入了一个实现了 Runnable 接口的对象。

源文件 11-8 FibonacciRunnableLambda.java

```
1 public class FibonacciRunnableLambda {
2     public static void main(String[] args) {
3         Thread t = new Thread(
4             () -> {
5                 long result = fib(7);
6                 System.out.println(result);
7             });
8
9         t.start();
10    }
11    private static long fib(long n) {
12        if (n == 0) {
13            return 0L;
14        }
15        if (n == 1) {
16            return 1L;
17        }
18        return fib(n - 1) + fib(n - 2);
19    }
20
21 }
```

11.3 线程状态

线程从被创建开始到终止总是处于这6个状态之一:新建(NEW)、可运行(RUNNABLE)、阻塞(BLOCKED)、等线程(WAITING)、等时间(TIMED _ WAITING)和终止

（TERMINATED）。线程的 6 个状态及其含义见表 11-1。

表 11-1 线程状态

状态	描述
新建（NEW）	线程对象已经被创建，但尚未启动
可运行（RUNNABLE）	线程一切就绪，随时可以被调度到中央处理器上运行
阻塞（BLOCKED）	由于访问的资源被锁住而处于不可运行状态
等线程（WAITING）	无限等待另外一个线程执行特定的动作
等时间（TIMED_WAITING）	线程以参数指定毫秒数处于不可运行状态
终止（TERMINATED）	线程停止运行，不再被调度

一个比较典型的线程的状态转移如图 11-3 所示。一旦线程被创建，就处于新建 NEW 状态。在 NEW 状态下，线程仅仅是内存中的一个对象；NEW 状态的线程执行 start 方法启动线程，线程转移到 RUNNABLE 状态，RUNNABLE 状态的线程随时准备在中央处理器上运行。

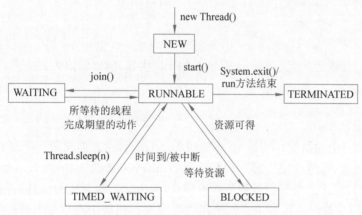

图 11-3 线程的状态的典型转移

处于 RUNNABLE 状态、BLOCKED 状态、WAITING 状态和 TIME_WAITING 状态之一的线程都是"活着的"线程。线程的实例方法 isAlive 就返回线程对象是否活着。

让另一线程执行无参数的 join() 方法后的线程进入 WAITING 状态，直到执行 join() 方法的线程终止后才继续执行；线程调用了 Thread.sleep(n) 后进入 TIME_WAITING 状态，当参数指定的时长（n 毫秒）到，线程回到 RUNNABLE 状态；当线程访问共享资源，需要进行同步/互斥时，可能进入 BLOCKED 状态，一旦资源可得，返回 RUNNABLE 状态。

线程有三种方式进入终止（TERMINATED）状态：run 方法执行结束；线程执行了 System.exit() 方法；线程抛出异常（Exception）或错误（Error）。

源文件 11-9 验证了 NEW 状态和 RUNNALE 状态之间的转移以及实例方法 isAlive() 的用法。其中线程的实例方法 getState() 返回枚举类型 Thread.State，其取值是 Thread 类中定义的枚举量：NEW、BLOCKED、RUNNABLE、TERMINATED、WAITING、TIMED_WAITING。

源文件 11-9　TestAliveThread.java

```
1 public class TestAliveThread {
2     public static void main(String[] argv) {
3
4         Thread a = new ThreadA();
5         System.out.println("State: " + a.getState());
6         System.out.println("is alive: " + a.isAlive());
7
8         a.start();
9         System.out.println("State: " + a.getState());
10        System.out.println("is alive: " + a.isAlive());
11    }
12 }
13
14 class ThreadA extends Thread {
15     public void run() {
16         System.out.println("I am ThreadA.");
17     }
18 }
```

在第 4 行创建线程后线程处于 NEW 状态，所以输出为：

```
State: NEW
is alive: false
```

在第 8 行启动线程后，线程处于 RUNNABLE 状态，所以输出为：

```
State: RUNNABLE
is alive: true
I am ThreadA.
```

源文件 11-10 中的第 3 行到第 13 行使用 Lambda 表达式创建了一个线程，这个线程一旦运行就进入休眠，计划休眠 7 秒。但是当运行 main 方法的 main 线程启动该线程后立即向其发出中断信号，这个中断信号被该线程捕获，输出运行时刻栈的信息后使用 break 语句退出 while 循环，也就终止运行了。如果在线程中没有捕获中断信号 InterruptedException 的语句，那么线程将一直休眠而整个程序则不会终止。

源文件 11-10　TestSleepThread.java

```
1 public class TestSleepThread{
2     public static void main(String[] args) {
3         Thread t = new Thread(
4             () -> {
5                 while (true) {
6                     try {
7                         Thread.sleep(7000);
8                     } catch (InterruptedException e) {
9                         e.printStackTrace();
10                        break;
11                    }
12                }
13            });
14        t.start();
```

```
15        System.out.println("main 线程发出中断信号");
16        t.interrupt();
17    }
18 }
```

系统线程,比如垃圾收集线程,称为守护(daemon)线程。通常使用守护线程作为后台线程完成侦听、服务或计时等任务。静态方法 Thread.setDaemon()用于将线程设置为守护线程。

如果需要等待另一个线程运行终止,则调用该线程的实例方法 join()使自己进入 WAITING 状态来实现。在源文件 11-11 中第 19 行创建了名为 t1 的线程,该线程一旦启动就休眠 7 秒。主线程启动 t1 后立即向 t1 发出 join 消息,通知 t1 自己在等它。所以直到 t1 在 7 秒后终止运行,主线程才继续运行,执行第 22 行的输出语句。因此程序输出为:

```
t1 is finished.
main thread is finished.
```

源文件 11-11　TestJoinThread.java

```
1 public class TestJoinThread {
2     public static class T extends Thread {
3         public T(String name) {
4             super(name);
5         }
6
7         @Override
8         public void run() {
9             try {
10                Thread.sleep(7000);
11            } catch (InterruptedException e) {
12                e.printStackTrace();
13            }
14            System.out.println(this.getName() + " is finished.");
15        }
16    }
17
18    public static void main(String[] args) throws InterruptedException {
19        T t1 = new T("t1");
20        t1.start();
21        t1.join();
22        System.out.println("main thread is finished. ");
23    }
24 }
```

如果没有第 21 行 t1.join(),程序的输出为:

```
main thread is finished.
t1 is finished.
```

从处于可运行(RUNNABLE)状态的线程中选择一个让其在中央处理器上执行的过程称为线程调度(scheduling)。Java 使用基于优先级的抢先调度策略(priority-based, preemptive)。

Java 中的每个线程都有优先级属性。优先级是从 1 到 10 的整数之一。使用实例方法

setPriority 可以改变优先级。Thread 类中定义了三个优先级常量：

```
MIN_PRIORITY = 1
NORM_PRIORITY = 5
MAX_PRIORITY = 10
```

线程的实例方法 getPriority 返回线程的当前优先级。默认优先级是 5。任何时刻，Java 的运行时刻系统总是从处于 RUNNABLE 状态的线程中选择优先级最高的线程运行。这就是基于优先级的线程调度思想。当高优先级线程运行期间又有更高优先级的称为 RUNNABLE 状态，Java 运行时刻系统就会中断当前线程的执行，立即运行这个新的更高优先级的线程。这就是"抢先"的思想。在分时操作系统中，Java 线程的调度会受到时间片的影响。

在源文件 11-12 中的第 18 行设置计算第 40 个 Fibonacci 数的线程的优先级最低；在第 23 行设置计算绝对值的线程的优先级最高。这样即便计算绝对值的线程后启动，也是先运行。程序的输出为：

```
主线程开始执行...
主线程终止.
子线程执行任务 2: 50
子线程执行任务 1: 102334155
```

源文件 11-12　TestThreadPriority.java

```
1  public class TestThreadPriority {
2      private static long fib(long n) {
3          if (n == 0) {
4              return 0L;
5          }
6          if (n == 1) {
7              return 1L;
8          }
9          return fib(n - 1) + fib(n - 2);
10     }
11
12     public static void main(String[] args) throws InterruptedException {
13         System.out.println("主线程开始执行...");
14
15         Thread threadFib = new Thread(() -> {
16             System.out.println("子线程执行任务 1: " +  fib(40));
17         });
18         threadFib.setPriority(Thread.MIN_PRIORITY);
19
20          Thread threadAbs = new Thread(() -> {
21             System.out.println("子线程执行任务 2: " + Math.abs(-50));
22          });
23         threadAbs.setPriority(Thread.MAX_PRIORITY);
24
25         threadFib.start();
26         threadAbs.start();
27         System.out.println("主线程终止.");
28     }
29 }
```

11.4 线程池

在工程实践中，建议使用线程池来运行线程。一个完成特定功能的方法称为一个任务(Task)，例如计算 Fibonacci 序列中的第 n 个数，创建线程就是把任务交给线程池中的一个空闲的线程运行。线程池预先创建好了若干线程，如果有任务到来就直接分配一个空闲线程来执行该任务，用完之后不是直接将其关闭，而是将其返回到线程池中，给其他任务使用。这样节省了创建和销毁线程的时间，提升了系统的性能。

java.util.concurrent.ThreadPoolExecutor 是执行任务的线程池。如果线程池中当前执行任务的线程数小于 corePoolSize，那么当新的任务请求执行时立即进入线程池；如果线程池中当前正在执行任务的线程数目大于或等于 corePoolSize，那么当新的任务被提交后进入等待队列；如果线程池中线程数目超过 maximumPoolSize，则抛出异常，拒绝执行该任务。构造 ThreadPoolExecutor 对象至少需要 4 个参数：

```
ThreadPoolExecutor(int corePoolSize, int maximumPoolSize, long keepAliveTime,
TimeUnit unit, BlockingQueue<Runnable> workQueue)
```

其中，当一个任务被提交到线程池时，线程池会分配一个线程来执行它，直到执行任务的线程数大于 corePoolSize。

maximumPoolSize 是线程池中能够持有的线程数目，即线程池的容量。如果等待队列满并且已经创建的线程数小于 maximumPoolSize，则线程池会再创建新的线程执行新提交的任务。如果等待队列无界，当执行任务的线程数大于 corePoolSize 时所有新提交的任务都会入队，这个参数失效。

keepAliveTime：线程池中空闲(idle)线程的最大的空闲时长。空闲线程指完成了任务的线程。当处理大量小任务时，调大这个时间能提高线程的利用率，提高线程池的吞吐量。

unit：keepAliveTime 的单位，有天、时、分、秒、毫秒、微秒和纳秒，是一个枚举量(java.util.concurrent.TimeUnit)。

workQueue：等待进入线程池的任务队列。可使用的队列有 ArrayBlockingQueue、LinkedBlockingQueue、SynchronousQueue、PriorityBlockingQueue 等。

可选的参数如下。

threadFactory：自定义创建线程的工厂。

handler：饱和策略，当线程池无法接纳新提交的任务时，按照饱和策略处理提交的新任务。JDK 提供 4 种策略：抛出异常放弃执行(AbortPolicy)、调用者处理(CallerRunsPolicy)、出队一个任务执行当前任务(DiscardOldestPolicy)和直接丢弃(DiscardPolicy)。

使用线程池的一般过程是：创建线程池对象、提交任务、关闭线程池。

调用线程池的 execute 方法来提交无返回值的任务，execute 方法的执行过程如下。

(1) 判断线程池中运行的线程数是否小于 corePoolSize。如果是，则安排一个线程来处理任务；否则执行下一步。

(2) 把任务添加到等待队列 workQueue 中。如果队列已满，则进入下一步。

(3) 判断线程池中运行的线程数是否小于 maximumPoolSize。如果是，则新建线程处理当前提交的任务；否则拒绝新任务，使用 handler 执行被拒绝的任务。

有 2 个方法关闭线程池：shutdownNow 和 shutdown。无论哪个方法，线程池给所有线程发送中断信号，线程响应中断信号终止运行。当所有线程都关闭之后，线程池的 isTerminaed() 方法调用返回 true。调用 shutdown 方法之后，线程池将不再接收新任务，所有已提交的任务处理完毕后，执行该任务的线程立刻终止；而调用 shutdownNow 方法后，线程池会将等待队列清空，所有线程执行完任务后立即终止。大多数情况下应使用 shutdown 方法来关闭线程池。

ThreadPoolExecutor 在 JDK 类层次中的位置如图 11-4 所示。

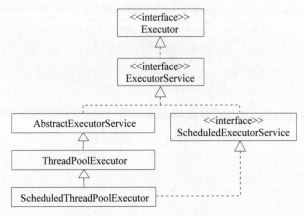

图 11-4 ThreadPoolExecutor

DemoThreads 演示了使用线程池来运行 10 个任务，每个任务都计算第 46 个 Fibonacci 数。

源文件 11-13 ThreadPoolExecutorDemo.java

```
 1 import java.util.concurrent.ArrayBlockingQueue;
 2 import java.util.concurrent.Executors;
 3 import java.util.concurrent.ThreadPoolExecutor;
 4 import java.util.concurrent.TimeUnit;
 5
 6 public class ThreadPoolExecutorDemo {
 7     static ThreadPoolExecutor threadPool = new ThreadPoolExecutor(3,
 8         5,
 9         10,
10         TimeUnit.SECONDS,
11         new ArrayBlockingQueue<Runnable>(10)
12     );
13     public static void main(String[] args) throws InterruptedException{
14         for (int i = 1; i <= 10; i++) {
15             String taskName = "Task" + i;
16             threadPool.execute(() -> {
17                 System.out.println(Thread.currentThread().getName()
18                     + " " + taskName + ": " + fib(46));
19             });
20         }
21         threadPool.shutdown();
```

```
22          System.out.println(Thread.currentThread().getName()
23              + " is over. ");
24      }
25
26      private static long fib(long n) {
27          if (n == 0) {
28              return 0L;
29          }
30          if (n == 1) {
31              return 1L;
32          }
33          return fib(n - 1) + fib(n - 2);
34      }
35 }
```

某次运行的结果是：

```
pool-1-thread-1 Task1: 1836311903
pool-1-thread-3 Task3: 1836311903
pool-1-thread-2 Task2: 1836311903
pool-1-thread-1 Task4: 1836311903
pool-1-thread-2 Task6: 1836311903
pool-1-thread-3 Task5: 1836311903
pool-1-thread-1 Task7: 1836311903
pool-1-thread-2 Task8: 1836311903
pool-1-thread-3 Task9: 1836311903
pool-1-thread-1 Task10: 1836311903
```

从运行结果可以看到，因为 corePoolSize 为 3，maximumPoolSize 为 5，等待队列容量为 10，任务数为 10，所以线程池使用了 pool-1-thread-1、pool-1-thread-2、pool-1-thread-3 三个线程先分别把 Task1、Task2 和 Task3 执行完毕，其他 7 个任务入队；然后从任务队列中分别把 Task4、Task5 和 Task6 出队并执行；再从任务队列中分别把 Task7、Task8 和 Task9 出队并执行；最后由 pool-1-thread-1 执行 Task10。

如果把第 21 行的 shutdown() 改为 shutdownNow()，则某次运行的结果为：

```
pool-1-thread-3 Task3: 1836311903
pool-1-thread-2 Task2: 1836311903
pool-1-thread-1 Task1: 1836311903
```

第 22 行的期望输出是在所有子线程结束输出后主线程才输出 "main is over. "：

```
pool-1-thread-3 Task3: 1836311903
pool-1-thread-2 Task2: 1836311903
pool-1-thread-1 Task1: 1836311903
pool-1-thread-3 Task4: 1836311903
pool-1-thread-1 Task6: 1836311903
pool-1-thread-2 Task5: 1836311903
pool-1-thread-2 Task9: 1836311903
pool-1-thread-3 Task7: 1836311903
pool-1-thread-1 Task8: 1836311903
pool-1-thread-2 Task10: 1836311903
main is over.
```

实际的输出反倒是"main is over."最先：

```
main is over.
pool-1-thread-3 Task3: 1836311903
pool-1-thread-2 Task2: 1836311903
pool-1-thread-1 Task1: 1836311903
pool-1-thread-3 Task4: 1836311903
pool-1-thread-1 Task6: 1836311903
pool-1-thread-2 Task5: 1836311903
pool-1-thread-2 Task9: 1836311903
pool-1-thread-3 Task7: 1836311903
pool-1-thread-1 Task8: 1836311903
pool-1-thread-2 Task10: 1836311903
```

这是因为主线程和10个子线程一样处于可运行（RUNNABLE）状态，而主线程安排好子线程运行后就没事了。

要想让主线程在所有10个子线程结束后再执行输出"main is over."的语句，则可以使用 CountDownLatch。CountDownLatch 让线程一直等，直到其他线程执行完自己的任务。用法是让每个线程在任务的末尾调用其实例方法 countDown，报告自己执行完任务了；线程的等待者则调用其实例方法 await 等待所有线程完成任务。创建 CountDownLatch 对象时需要指定任务数。

当调用 CountDownLatch 对象的实例方法 countDown 时，任务数会被减1；当调用 await 方法时，判断当前任务数是否大于 0。是，则线程会被阻塞，一直到任务数被 countDown 方法减到 0 时，线程才会继续执行下一条语句。源文件 11-14 中的类 DemoCountDownLatch 就能够输出预期的结果。

源文件 11-14　CountDownLatchDemo.java

```
 1 import java.util.concurrent.ArrayBlockingQueue;
 2 import java.util.concurrent.ThreadPoolExecutor;
 3 import java.util.concurrent.TimeUnit;
 4 import java.util.concurrent.CountDownLatch;
 5
 6 public class CountDownLatchDemo   {
 7     static final int TASK_COUNT = 10;      //任务总数
 8
 9     static ThreadPoolExecutor threadPool = new ThreadPoolExecutor(3,
10         5,
11         10,
12         TimeUnit.SECONDS,
13         new ArrayBlockingQueue<Runnable>(TASK_COUNT)
14     );
15     //单次计数器
16     static CountDownLatch countDownLatch = new CountDownLatch(TASK_COUNT);
17     public static void main(String[] args) throws InterruptedException {
18         for (int i = 1; i <= TASK_COUNT; i++)   {
19             String taskName = "Task" + i;
20             threadPool.execute(() -> {
21                 System.out.println(Thread.currentThread().getName()
22                     + " " + taskName + ": " + fib(46));
```

```
23                    //线程执行完,计数器-1
24                    countDownLatch.countDown();
25                });
26            }
27            //阻塞,等待线程池任务执行完
28            countDownLatch.await();
29            threadPool.shutdown();
30            System.out.println(Thread.currentThread().getName()
31                + " is over. ");
32        }
33
34        private static long fib(long n) {
35            if (n == 0) {
36                return 0L;
37            }
38            if (n == 1) {
39                return 1L;
40            }
41            return fib(n - 1) + fib(n - 2);
42        }
43    }
```

第 16 行声明一个包含了 10 个任务的计数器;第 24 行让每个任务执行完之后任务数-1;第 28 行让主线程阻塞,等待 CountDownLatch 对象把任务数减为 0,表示任务都执行完了,可以继续执行 await 方法后面的语句了。

11.5 线程安全的程序设计

11.5.1 与时间有关的错误

当多个线程访问同一个对象,如果没有任何机制加以协调和控制,那么就会产生与时间有关的错误。下面的类 Counter 展示了这个问题。

```
class Counter {
    private int count = 1;
    public void increment() {
        count = count + 1;
    }
    public void decrement() {
        count = count - 1;
    }
    public int getValue() {
        return count;
    }
}
```

这个 Counter 类用于创建被多个线程访问的共享对象。Counter 类的实例方法 increment 把私有成员变量 count 增 1;而实例方法 decrement 则把私有成员变量 count 减 1。

由于线程何时被调度运行、能够运行多长时间都是不确定的,所以当多线程访问 Counter 对象时,就可能产生与时间有关的错误。

假设线程 A 让 Counter 对象执行 increment 方法;线程 B 让 Counter 对象也执行 increment 方法;并假设私有成员变量 count 的当前值是 1。

那么线程 A 和线程 B 一种可能的交替执行序列是:

Thread A:读取变量 count 的当前值到寄存器。

Thread B:读取变量 count 的当前值到寄存器。

Thread A:在寄存器中把当前值增 1,结果是 2。

Thread B:在寄存器中把当前值增 1,结果是 2。

Thread A:把计算结果 2 存入变量 count。

Thread B:把计算结果 2 存入变量 count。

这种交替执行的结果是:Counter 对象的私有成员变量 count 的值经过线程 A 增 1 和线程 B 增 1 变成了 2;而不是所期望的 3。也有可能没有问题,count 的最后的值是期望的 3。

下面是一个共享 Counter 对象的应用。为了控制进入景区的游客数量,景区在各个入口统计进入和离开景区的游客。假设某景区有两个入口:1 号门和 2 号门。进入一名游客计数器增 1;离开一名游客计数器减 1。假设 3 名游客从 1 号门进入后由 2 号门离开。每名进入游客对应一个线程,每名离开游客对应一个线程。监控景区当前游客数目的程序如源文件 11-15 所示。

源文件 11-15 ThreadInterference.java

```java
1 import java.util.concurrent.ArrayBlockingQueue;
2 import java.util.concurrent.ThreadPoolExecutor;
3 import java.util.concurrent.TimeUnit;
4 /**
5  * A counter to count the current number of visitors in a scenic area.
6  * Suppose there are 2 entries. In order to limit the number of
7  * visitors, guards needs to know the current number of visitors.
8  * Each entry runs a program to increament or decrement the number.
9  */
10
11 class Counter {
12     int count = 0;
13
14     int increment() {
15         count = count + 1;
16         return count;
17     }
18
19     int decrement() {
20         if (count > 0) {
21             count = count - 1;
22         }
23         return count;
24     }
25
26     int getCount() {
27         return count;
```

```
28       }
29 }
30
31 public class ThreadInterference {
32     public static void main(String[] args) {
33         final Counter c = new Counter();
34         long starting = System.nanoTime();
35         ThreadPoolExecutor executor = new ThreadPoolExecutor(
36             10,
37             10,
38             0,
39             TimeUnit.SECONDS,
40             new ArrayBlockingQueue<Runnable>(16));
41
42         for (int i = 1;  i <= 3;   i++) {
43             executor.execute(
44                 new Thread(() -> {
45                     System.out.println("Gate 1: " + c.increment()
46                         + " at " + (System.nanoTime() - starting));
47                 }));
48         }
49
50         for (int i = 1; i <= 3; i++) {
51             executor.execute(
52                 new Thread(() -> {
53                     System.out.println("Gate 2: " + c.decrement()
54                         + " at " + (System.nanoTime() - starting));
55                 }));
56         }
57         executor.shutdown();
58         //Thread.sleep(1000);
59     }
60 }
```

某次运行的输出是：

```
Gate 2: 0 at 6376800
Gate 2: 0 at 5919300
Gate 1: 1 at 5791600
Gate 2: 1 at 5869400
Gate 1: 2 at 5844900
Gate 1: 1 at 5931700
```

把该输出按线程运行的时刻排序：

```
Gate 1: 1 at 5791600
Gate 1: 2 at 5844900
Gate 2: 1 at 5869400
Gate 2: 0 at 5919300
Gate 1: 1 at 5931700
Gate 2: 0 at 6376800
```

计数器应该先增加到 3,然后再减少到 0,期望的次序是:

```
Gate 1: 1
Gate 1: 2
Gate 1: 3
Gate 2: 2
Gate 2: 1
Gate 2: 0
```

在源文件 11-15 中,程序的设计意图是 1 号门连续进入 3 名游客;2 号门连续离开 3 名游客。但某次的实际运行结果是 1 号门进入 2 名;2 号门离开 2 名;1 号门进入 1 名;2 号门离开 1 名。这种多线程程序设计中产生的与期望行为不一致的错误与语句运行的时间相关,称为"与时间有关的错误"。不仅输出的次序与访问计数器的次序不同,也可能产生错误的输出。必须有一种机制保证多线程环境下不会出现与时间有关的错误。

不同的线程共享内存,但不共享寄存器。当执行 increament 方法时会把内存变量 count 的当前值读入寄存器,然后更新,再写回内存。在冯·诺依曼体系中,指令和数据存储在内存中,CPU 负责取指和执行,如图 11-5 所示。

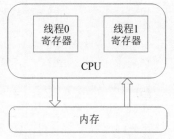

图 11-5　线程的运行环境

从图 11-5 中可以看出,如果线程 A 和线程 B 共享内存中的变量,那么线程 A 把寄存器中更新过的值写回到内存中去;线程 B 到主内存中去读取线程 A 之前已更新过的共享变量。

JVM 不保证实例变量 count 的访问是原子的,有可能不同线程读入到寄存器、更新、写回内存的指令交叠执行,造成与时间有关的错误。如果线程 A 修改了内存中的共享变量之后,不被中断地将其写回内存;线程 B 每次读共享变量时,都去内存中重新读取到寄存器,就会避免与时间有关的错误。线程没有看到变量的最新值,因为它还没有被另一个线程写回主存,这种问题被称为"可见性(visibility)"问题。

线程安全的程序设计是指在多个线程同时访问共享资源的情况下,程序仍然产生正确的和可预期的输出。以下是实现线程安全程序设计的需求。

(1) 可见性(visibility)。确保当一个线程修改了共享资源后,其他线程能够立即看到这些修改。通常使用 volatile 保留字实现。

(2) 原子性(atomicity)。如果一个对共享资源的操作包含若干动作,那么这些动作要么全部执行完毕;要么一个也不执行。确保在任意时刻只有一个线程能够原子地操作共享资源。通常使用 synchronized 关键字或显式锁来实现。

(3) 避免死锁(deadlock avoidance)。防止出现线程互相等待对方释放资源而导致的所有线程都无法继续执行的情况。

11.5.2　volatile 保留字

用 volatile 修饰变量在对其写和读之间建立 happens-before 关系:在任意线程"读"之前,某线程的"写"已经完成。happens-before 是语句块在线程间的偏序关系;某语句块的执行必须在另外一语句块执行之前发生,无论这两个语句块在哪个线程中。被 volatile 修饰的变量能确保所有线程看到的是同一个值。

源文件 11-16　TestVolatile.java

```java
1 class Counter {
2      /**
3       * 保留字 volatile 确保一个线程造成的变化会立即反应给其他线程
4       *
5       */
6      static volatile int count = 0;
7
8      int increment() {
9          count = count + 1;
10         return count;
11     }
12
13     int decrement() {
14         if (count > 0) {
15             count = count - 1;
16         }
17         return count;
18     }
19
20     int getCount() {
21         return count;
22     }
23 }
24
25 public class TestVolatile {
26     public static void main(String[] args) {
27         final Counter c = new Counter();
28
29         Thread t1 = new Thread(() ->  {
30             for (int i = 1; i <= 3; i++) {
31                 System.out.println("Gate   1: " + c.increment()
32                     + " at " + System.nanoTime());
33             }
34         });
35
36         Thread t2 = new Thread(() ->  {
37             for (int i = 1; i <= 3; i++) {
38                 System.out.println("Monitor: " + c.getCount()
39                     + " at " + System.nanoTime());
40             }
41         });
42
43         t1.start();
44         t2.start();
45     }
46 }
```

源文件 11-16 中创建了两个线程：线程 t1 用于修改变量 count；另一个线程 t2 用于读变量 count。成员变量 count 前的 volatile 保留字保证了 t2 总能够正确读出最新的 count 值。

例如当线程 t1 把变量 count 增 1 后，另外一个线程 t2 仅读取 count，不修改，这种情形下，把 count 声明为 volatile 可以保证 t2 能够看到 t1 的写。源文件 11-16 中的第 43 行启动线程，让其修改共享变量 count；接着在第 44 行启动另外一个线程读取一次。程序的输出结果如下：

```
Monitor: 0 at 622784086133400
Monitor: 1 at 622784095185900
Monitor: 1 at 622784095208600
Gate 1: 1 at 622784086425200
```

从输出结果看到，读 count 的线程先运行，读出的是 0；写 count 的线程接着运行，写入 1；后面读线程又读了 2 次，分别输出 1。不会出现写入 1 而读出 0 的情况。

11.5.3 synchronized 保留字

如果两个线程都在修改共享变量，那么仅仅使用 volatile 关键字是不够的。比如线程 A 执行 increament，刚在寄存器中把当前值增 1，尚未写回内存而退出执行；线程 B 开始执行，当线程 B 并不知道线程的工作进度，只顾自己从内存中读、在自己的寄存器中更新、写回内存。本来被线程 A 和 B 两次增 1，但是线程 A 的增 1 被丢失了。这种多线程更新的情况下，需要使用保留字 synchronized 来保证变量的读写是原子的。

源文件 11-15 展示的与时间有关的错误可通过关键字 synchronized 解决。景区有两个门均可出入：Gate 1 和 Gate 2。无论游客从哪个门进入，哪个门离开，都得更新计数器。源文件 11-17 第 42 行的循环语句让线程池执行 3 个游客进入的任务；第 53 行的循环语句让线程池执行 3 个游客离开的任务。任务所调用的实例方法 increment()、decrement() 都是在第 9 行和第 14 行 synchronized 修饰的同步方法。关键字 synchronized 使得一个线程在执行 Counter 的实例方法 increment() 期间禁止其他线程执行这个方法。这样，任何时刻只允许一个线程执行 increment() 方法。如果线程 A 执行 increment 方法期间线程 B 也要执行 increment() 方法，那么线程 B 就会转入 "BLOCKED" 状态，直到线程 A 结束 increment() 方法的执行。

源文件 11-17　SynchronizedThread.java

```
1 import java.util.concurrent.ArrayBlockingQueue;
2 import java.util.concurrent.ThreadPoolExecutor;
3 import java.util.concurrent.TimeUnit;
4 import java.util.concurrent.CountDownLatch;
5
6 class Counter {
7     static int count = 0;
8
9     synchronized int increment() {
10        count = count + 1;
11        return count;
12    }
```

```java
13
14      synchronized int decrement() {
15          if (count > 0) {
16              count = count - 1;
17          }
18          return count;
19      }
20
21      synchronized int getCount() {
22          return count;
23      }
24  }
25
26  public class SynchronizedThread {
27      static final int TASK_COUNT = 6;      //任务总数
28      static CountDownLatch countDownLatch
29          = new CountDownLatch(TASK_COUNT);
30
31      public static void main(String[] args) throws InterruptedException {
32          final Counter c = new Counter();
33          long starting = System.nanoTime();
34
35          ThreadPoolExecutor executor = new ThreadPoolExecutor(
36              10,
37              10,
38              0,
39              TimeUnit.SECONDS,
40              new ArrayBlockingQueue<Runnable>(16));
41
42          for (int i = 1; i <= 3; i++) {
43              executor.execute(
44                  new Thread(
45                      () -> {
46                          c.increment();
47                          //线程执行完,计数器 -1
48                          countDownLatch.countDown();
49                      })
50              );
51          }
52
53          for (int i = 1; i <= 3; i++) {
54              executor.execute(
55                  new Thread(
56                      () -> {
57                          c.decrement();
58                          //线程执行完,计数器 -1
59                          countDownLatch.countDown();
60                      })
61              );
62          }
63          //阻塞,等待线程池任务执行完
64          countDownLatch.await();
65          executor.shutdown();
66          System.out.println("count: " + c.getCount());
67      }
68  }
```

如果一个对象有多个 synchronized 方法，某一时刻某个线程已经进入到了某个 synchronized 方法，那么在该方法没有执行完毕前，其他线程是无法访问该对象的任何 synchronized 方法的。

实际上这种线程间的同步是应用"锁"完成的：当线程 A 访问某个对象的 synchronized 方法时，将该对象上锁，试图访问该对象的任何一个 synchronized 方法线程 B 就要进入"BLOCKED"状态。这种状态直到线程 A 的访问的同步方法执行完毕（或者是抛出了异常），才将该对象上的锁释放掉，线程 B 从"BLOCKED"状态转回"RUNNABLE"状态，等待调度，重新访问该对象的 synchronized 方法。

与共享资源有关的所有方法都应声明为同步方法。当一个 synchronized 关键字修饰的方法同时又被 static 关键字修饰，这就是同步静态方法的情形。前文介绍过，同步实例方法会将对象上锁。但是静态方法属于类，不属于对象，所以同步静态方法的执行会给"类"上锁。因此，当线程 A 访问某类的静态同步方法时，如果线程 B 也试图访问该类任一静态同步方法，那么线程 B 就被阻塞，直到线程 A 把该静态同步方法执行完毕。

如果一个类中定义了一个同步静态方法 M，也定义了一个同步实例方法 N，那么线程 A 访问这个类同步静态方法 M，线程 B 访问该类对象的同步实例方法 N，线程 A 和线程 B 互不影响，因为它们的锁都不一样。M 方法的锁是"类"，N 锁是对象。

可以约束多个线程不能同时执行一个代码块。即在代码块（block）上进行线程同步，称为"同步块"。

同步块的语法形式如下：

```
synchronized(<对象>){
    ...
}
```

意思是当线程试图执行代码块时，先把<对象>上锁。这个对象可以是任意类的对象，当然也可以是当前对象 this。synchronized 方法实际上等同于用一个 synchronized 块包住方法中的所有语句，然后在 synchronized 块的括号中传入 this 关键字。当两个并发线程访问 synchronized(<对象>)同步代码块时，某时刻只能有一个线程得到执行。另一个线程必须等待当前线程执行完这个代码块以后才能执行该代码块。当一个线程访问 synchronized(<对象>)同步代码块时，它就为这个<对象>上锁。结果，其他试图为该<对象>上锁线程都被暂时阻塞。

源文件 11-18　SynchronizedBlock.java

```
1  class Task implements Runnable {
2      @Override
3      public void run() {
4          synchronized (SynchronizedBlock.block) {
5              for (int i = 0; i < 300; i++) {
6                  SynchronizedBlock.count++;
7              }
8          }
9      }
10 }
11
12 public class SynchronizedBlock {
```

```
13      static Object block = new Object();
14      static int count = 0;
15
16      public static void main(String[] args)
17              throws InterruptedException {
18          Task t = new Task();
19          Thread t1 = new Thread(t, "t1");
20          Thread t2 = new Thread(t, "t2");
21          t1.start();
22          t2.start();
23          t1.join();
24          t2.join();
25          System.out.println("count: " + count);
26      }
27  }
```

源文件 11-18 首先创建了一个 Object 对象并使用静态变量 block 引用,以便被 Task 类访问。"任务"类 Task 实现 Runnable 接口,重复 300 次对静态变量 count 增 1 操作。在第 4 行使用关键字 synchronized 对静态引用变量 block 所引用的对象同步。当线程 t1 执行 for 循环时,先为对象 block 上锁;这样,如果线程 B 也试图执行 for 循环,就会被阻塞。

把第 4 行中的"SynchronizedBlock.block"换成当前对象 this,效果相同。同步方法是指对整个方法进行加锁同步,而同步块是指对方法内的某个代码块进行加锁同步。同步方法的锁用的是其实例对象本身,而同步代码块的锁可以自己指定。

11.5.4 计数器 Adder

java.util.concurrent.atomic.LongAdder 降低了高并发情况下锁的粒度,效率比较高。源文件 11-19 中第 8 行 new LongAdder() 创建一个 LongAdder 对象,初始值是 0,调用 increment() 方法可以对 LongAdder 内部的值原子递增 1。reset() 方法可以重置 LongAdder 的值,使其归 0。第 29 行的循环语句中启动了 10 个线程,每个线程对计数器 counter 增 1 共 12345 次,最终使得计数器的值为 12345×10=123450。程序控制这 10 个线程重复执行 3 次。每次运行结果都是 123450。

源文件 11-19 LongAdderDemo.java

```
1  import java.util.concurrent.atomic.LongAdder;
2  import java.util.concurrent.ExecutionException;
3  import java.util.concurrent.ArrayBlockingQueue;
4  import java.util.concurrent.ThreadPoolExecutor;
5  import java.util.concurrent.TimeUnit;
6
7  public class LongAdderDemo {
8      static LongAdder counter = new LongAdder();
9      static ThreadPoolExecutor executor = new ThreadPoolExecutor(
10         256,
11         256,
12         0,
13         TimeUnit.SECONDS,
14         new ArrayBlockingQueue<Runnable>(256));
15
```

```java
16    public static void main(String[] args)
17        throws ExecutionException, InterruptedException {
18        for (int i = 0; i < 3; i++) {
19            counter.reset();
20            start();
21            Thread.sleep(1000);
22            System.out.println(counter);
23        }
24        executor.shutdown();
25    }
26
27    private static void start() throws ExecutionException {
28        int threadCount = 10;
29        for (int i = 0; i < threadCount; i++) {
30            Thread t = new Thread(() -> {
31                for (int j = 0; j < 12345; j++) {
32                    counter.increment();
33                }
34            });
35            t.setName("Task" + i);
36            executor.execute(t);
37        }
38    }
39 }
```

LongAccumulator 是 LongAdder 的功能增强版。LongAdder 的 API 只有对数值的加减，而 LongAccumulator 提供了自定义的函数操作，其构造函数如下：

```
public LongAccumulator ( LongBinaryOperator accumulatorFunction, long identity) {
    this.function = accumulatorFunction;
    base = this.identity = identity;
}
```

其中 accumulatorFunction 是需要执行的二元累计函数，有两个 long 型形参，返回 1 个 long 型数值。identity 是初始值。

源文件 11-20 的第 10 行使用完成二元加法运算的匿名函数和初始值 0 创建了 LongAccumulator 替换了源文件 11-19 中的 LongAdder 对象，其他操作相同。其实调用 new LongAdder() 等价于 new LongAccumulator((x, y) -> x+y, 0L)。

源文件 11-20　LongAccumulatorDemo.java

```java
1 import java.util.concurrent.atomic.LongAccumulator;
2 import java.util.concurrent.ExecutionException;
3 import java.util.concurrent.ArrayBlockingQueue;
4 import java.util.concurrent.ThreadPoolExecutor;
5 import java.util.concurrent.TimeUnit;
6
7 public class LongAccumulatorDemo {
8    static LongAccumulator counter
9        = new LongAccumulator((x, y) -> x + y, 0L);
10   static ThreadPoolExecutor executor = new ThreadPoolExecutor(
11       256,
```

```
12          256,
13          0, 0
14          TimeUnit.SECONDS,
15          new ArrayBlockingQueue<Runnable>(256));
16
17      public static void main(String[] args)
18          throws ExecutionException, InterruptedException {
19          for (int i = 0; i < 3; i++) {
20              counter.reset();
21              start();
22              Thread.sleep(1000);
23              System.out.println(counter);
24          }
25          executor.shutdown();
26      }
27
28      private static void start() throws ExecutionException {
29          int threadCount = 10;
30          for (int i = 0; i < threadCount; i++) {
31              Thread t = new Thread(() -> {
32                  for (int j = 0; j < 12345; j++) {
33                      counter.accumulate(1);
34                  }
35              });
36              t.setName("Task" + i);
37              executor.execute(t);
38          }
39      }
40  }
```

11.6 获取子线程的返回结果

Runnable 接口的 run()方法的局限在于不能够返回值，或者抛出 checked 异常。当期望线程把计算结果返回时应让线程实现 Callable 接口而不是 Runnable 接口。与使用 Runnable 接口相比，Callable 的 call()方法可以有返回值，可以抛出异常。

java.util.concurrent.FutureTask 接收一个 Callable 或 Runnable 对象作为任务，创建线程执行该任务并记录任务的状态（等待、运行、完成、取消等），以及计算结果或异常信息，提供了判断任务是否已完成、获取任务的执行结果等实例方法。每个 FutureTask 只能执行一次，如果需要再次执行，需要创建新的 FutureTask 对象。

FutureTask 的用法如下。

(1) 创建 FutureTask 任务。以一个 Callable 对象为参数来创建 FutureTask 任务。

(2) 提交 FutureTask 任务给线程执行。

(3) 获取 FutureTask 任务的返回结果。某线程调用 FutureTask 任务 get()方法后将会阻塞，直到 FutureTask 任务终止并返回结果。

(4) 取消任务的执行。FutureTask 任务的 cancel()方法用来取消任务的执行。参数 true 表示尝试中断任务的执行，false 表示不中断。

源文件 11-21 通过实现 Callable 接口设计了实现类 Accumulator，该类的功能是把 1 到 100 的数累加起来，并把累加结果返回。在第 18 行以 Callable 实现类 Accumulator 的实例为参数创建了 FutureTask<Integer> 对象，第 19 行以 FutureTask<Integer> 对象为任务创建线程并启动运行。通过 FutureTask<Integer> 对象的 get 方法获取累加结果，get() 方法阻塞当前线程。当 FutureTask<Integer> 的 run() 方法执行完毕之后，get() 方法才会从阻塞中返回。

源文件 11-21　CallableDemo.java

```java
1 import java.util.concurrent.Callable;
2 import java.util.concurrent.FutureTask;
3
4 class Accumulator implements Callable<Integer> {
5     @Override
6     public Integer call() throws Exception {
7         int sum = 0;
8         for (int i = 1; i <= 100; i++) {
9             sum += i;
10         }
11         return sum;
12     }
13 }
14
15 public class CallableDemo {
16     public static void main(String[] args) throws Exception {
17         Accumulator a = new Accumulator();
18         FutureTask<Integer> task = new FutureTask<Integer>(a);
19         new Thread(task).start();
20         System.out.println(task.get());
21     }
22 }
```

11.7　BlockingQueue

同步方法仅能保证对单一方法的原子访问。当需要"原子"地执行多个方法时，这种同步设施可能会失败。例如，假设有容器 Collection 对象 c，先判断里面有没有指定的元素，如果没有再添加元素：

```
Collection c;
if (!c.contains(element)){
    c.add(element);
}
```

即使 contains() 方法和 add() 方法都是同步的，也不能保证容器对象 c 被多线程正确地访问。

以队列为例，接口 java.util.concurrent.BlockingQueue 提供了线程安全的队列方案。该接口的实现类有 ArrayBlockingQueue、LinkedBlockingQueue 等。

源文件 11-22 展示了如何使用 BlockingQueue 解决生产者和消费者问题。生产者线程负责生产产品(本例中是一个整数对象),消费者负责消费产品(本例仅仅是把整数对象输出)。生产者线程把产品放入一个队列中,当队列无空间可用时则进入阻塞状态;消费者线程从队列中取走产品进行消费,当无产品可取时则进入阻塞状态。

表 11-2 BlockingQueue 中的方法

方 法	描 述
void put(E o)	入队,如果无可用空间则阻塞
E take()	出队,如果无元素可出队则阻塞

在源文件 11-22 第 6 行创建了一个 BlockingQueue 对象,在第 10 行创建了一个生产者线程对象,第 12 行和第 13 行创建了两个消费者线程对象。然后启动这三个线程。出于演示目的,只让生产者线程 ProducerThread 生产 10 个产品,所以定义了常量 MAX 等于 10,并在第 24 行定义变量 i 为计数器,在第 43 行判断是否已经生产了 10 件产品,如果是则返回 null。返回空会在第 29 行产生空指针异常,该异常被捕获后输出信息 Produce nothing,并退出循环。而消费者线程 ConsumerThread 没有退出循环的语句,所以程序一直运行。

源文件 11-22 BlockingQueueDemo

```
1  import java.util.concurrent.ArrayBlockingQueue;
2  import java.util.concurrent.BlockingQueue;
3
4  public class BlockingQueueDemo {
5      static final int MAX = 10, CAPACITY = 5;
6      static BlockingQueue<Integer> queue
7          = new ArrayBlockingQueue<Integer>(CAPACITY);
8
9      public static void main(String[] args) {
10         ProducerThread p = new ProducerThread();
11         p.setName("producer");
12         ConsumerThread c1 = new ConsumerThread();
13         ConsumerThread c2 = new ConsumerThread();
14         c1.setName("c1");
15         c2.setName("c2");
16
17         p.start();
18         c1.start();
19         c2.start();
20     }
21 }
22
23 class ProducerThread extends Thread {
24     Integer i = 0;
25     @Override
26     public void run() {
27         while (true) {
28             try {
29                 Integer i = produce();
30                 System.out.println("p: " + i);
31                 BlockingQueueDemo.queue.put(i);
32             } catch (InterruptedException e) {
```

```
33                e.printStackTrace();
34                break;
35            } catch (NullPointerException e) {
36                System.out.println("无.");
37                break;
38            }
39        }
40    }
41
42    private Integer produce() {
43        if (i == BlockingQueueDemo.MAX) {
44            return null; //将会导致异常: NullPointerException
45        } else {
46            return i++;
47        }
48    }
49 }
50
51 class ConsumerThread extends Thread {
52    @Override
53    public void run() {
54        while (true) {
55            try {
56                consume(BlockingQueueDemo.queue.take());
57            } catch (InterruptedException ex) {
58                System.out.println("阻塞期间被中断.");
59                break;
60            }
61        }
62    }
63
64    private void consume(Integer x) {
65        System.out.println(Thread.currentThread().getName()
66            + ":" + x);
67    }
68 }
```

某次运行的输出是：

```
p: 0
p: 1
p: 2
p: 3
p: 4
p: 5
p: 6
p: 7
c1:0
c1:2
c1:3
c1:4
c1:5
c1:6
p: 8
```

```
p: 9
p: null
Produce nothing.
c1:7
c1:8
c1:9
c2:1
```

从输出可以看到,生产者 p 生产了产品 0 到 7 后,消费者 c1 消费了 0,2,3,4,5,6;生产者又生产了产品 8 和 9,到达产品的上限不再生产,返回 null;消费者 c1 消费 7,8,9;消费者 c2 消费产品 1;最后两个消费者一直等待缓冲区中放入新产品。在 jGRASP 中需要单击 "End" 按钮结束运行。

源文件 11-23 使用线程池来运行生产者消费者线程。当主线程提交生产者线程和消费者线程给线程池运行后在第 24 行使用 sleep()方法休眠 1 秒,等待生产和消费了 10 个产品,然后在第 25 行调用线程池 shutdownNow()终止运行。然后某次运行的输出如下:

```
p: 0
p: 1
p: 2
p: 3
p: 4
p: 5
p: 6
p: 7
pool-1-thread-2: 0
pool-1-thread-3: 1
p: 8
pool-1-thread-2: 2
p: 9
pool-1-thread-3: 3
p: null
pool-1-thread-2: 4
pool-1-thread-3: 5
无.
pool-1-thread-2: 6
pool-1-thread-3: 7
pool-1-thread-2: 8
pool-1-thread-3: 9
消费者阻塞期间被中断.
消费者阻塞期间被中断.
```

在输出中,pool-1-thread-2 和 pool-1-thread-3 是线程池给消费者线程的名字。因为线程池 shutdownNow()抛出 InterruptedException 异常,两个消费者线程都捕获该异常,所以有两个"消费者阻塞期间被中断."输出。因为生产者线程在 shutdownNow()抛出 InterruptedException 异常之前已经由于空指针异常终止执行,所以只输出"无."。

源文件 11-23 BlockingQueuePool.java

```
1 import java.util.concurrent.ArrayBlockingQueue;
2 import java.util.concurrent.BlockingQueue;
3 import java.util.concurrent.ThreadPoolExecutor;
4 import java.util.concurrent.TimeUnit;
5
```

```java
 6 public class BlockingQueuePool {
 7     static final int MAX = 10, CAPACITY = 5;
 8     static BlockingQueue<Integer> queue
 9       = new ArrayBlockingQueue<Integer>(CAPACITY);
10     static ThreadPoolExecutor threadPool = new ThreadPoolExecutor(3,
11         5,
12         1,
13         TimeUnit.SECONDS,
14         new ArrayBlockingQueue<Runnable>(CAPACITY)
15     );
16
17     public static void main(String[] args) throws InterruptedException {
18         ProducerThread p = new ProducerThread();
19         ConsumerThread c1 = new ConsumerThread();
20         ConsumerThread c2 = new ConsumerThread();
21         threadPool.execute(p);
22         threadPool.execute(c1);
23         threadPool.execute(c2);
24         Thread.sleep(1000);
25         threadPool.shutdownNow();
26     }
27 }
28
29 class ProducerThread extends Thread {
30     Integer i = 0;
31     @Override
32     public void run() {
33         while (true) {
34             try {
35                 Integer i = produce();
36                 System.out.println("p: " + i);
37                 BlockingQueuePool.queue.put(i);
38             } catch (InterruptedException e) {
39                 e.printStackTrace();
40                 break;
41             } catch (NullPointerException e) {
42                 System.out.println("无.");
43                 break;
44             }
45         }
46     }
47
48     private Integer produce() {
49         if (i == BlockingQueuePool.MAX) {
50             return null; //导致 NullPointerException
51         } else {
52             return i++;
53         }
54     }
55 }
56
57 class ConsumerThread extends Thread {
58     @Override
```

```
59    public void run() {
60        while (true) {
61            try {
62                consume(BlockingQueuePool.queue.take());
63            } catch (InterruptedException ex) {
64                System.err.println("消费者阻塞期间被中断.");
65                break;
66            }
67        }
68    }
69
70    private void consume(Integer x) {
71        System.out.println(Thread.currentThread().getName()
72            + ": " + x);
73    }
74 }
```

第 11 章　章节测验

第 12 章 学生选课系统

12.1 需求分析

开发一个学生选课系统(CRS),实现大学生选修每学期开出的课程。

当大学生入学后,提供专业学习的学院一般已经为该年级的学生准备了四年的学习计划,即为了满足学历、学位的要求而应完成的课程(Course)。在每学期期末的选课(Register for Class)时间,学生都能通过选课系统查看下学期所开设的课(Class),选择和注册自己应该或有兴趣学习的课程。如果多名教师讲授这门课程,学生还可以选择自己喜欢的教师的课。如果有空余的选课名额,那么学生就选课成功。任课教师可以查看选课学生名单。

要求把学生选课系统设计为一个控制台应用,仅实现选课功能。程序启动后,首先输出由所有授课教师开出的课程和所有准备开出的课:

```
=========学生=========
张三 (201901001) [学士 - 数学]
李四 (201901002) [学士 - 计算机科学与技术]
王五 (201901003) [学士 - 计算机科学与技术]
=========教师=========
董永 (副教授, 计算机科学与技术)
赵云 (教授, 计算机科学与技术)
郭天 (教授, 数学)
=====课程 Course=====
CS103: 数据结构与算法, 3.0 学分
MAT101: 概率与统计, 3.0 学分
CS101: C程序设计, 3.0 学分
CS102: 面向对象程序设计, 3.0 学分
CS201: 离散数学, 3.0 学分
===开出的课 ScheduledCourse===
CS102-2, 周四, 下午 4:00-6:00, 郭天 (教授, 数学), 25
CS103-1, 周一, 下午 6:00-8:00, 董永 (副教授, 计算机科学与技术), 20
```

```
MAT101-1, 周五, 下午 4:00-6:00, 赵云 (教授, 计算机科学与技术), 15
CS101-1, 周一, 上午 8:00-10:00, 郭天 (教授, 数学), 30
CS101-2, 周二, 上午 8:00-10:00, 赵云 (教授, 计算机科学与技术), 30
CS102-1, 周三, 下午 2:00-4:00, 董永 (副教授, 计算机科学与技术), 25
CS201-1, 周一, 下午 4:00-6:00, 郭天 (教授, 数学), 1
```

在"开出的课"部分的最后一列是班容量。最后一行 CS201-1 的班容量是 1,即只有一个学生。然后演示一个学生注册成功,另外一个学生再注册该课则失败的例子:

```
学生 张三 试图注册 CS201-1
注册成功
学生 王五 试图注册 CS201-1
注册失败:已满员
```

教师和学生具有"人"的共同属性,如都有名字。所以把 Teacher 类和 Student 类作为 Person 类的子类。学生具有专业 major 和学位 degree 等属性,教师具有职称 title 等属性。课(class)是安排了资源的课程(Course),课的属性包括周几 dayOfWeek、上课时间 timeOfDay、教室 room 和班容量 seatingCapapcity 等。课程的属性有课程名 name 和学分 credits 等。为了避免和 Java 的保留字 class 混淆,这里使用 ScheduledCourse 代替 class。一个课程 Course 对象可以安排多个教学班,形成多个 ScheduledCourse 对象;一个教师可以教授多门课;一门课有一位任课教师。一名学生可以注册多门课,一门课里有多名注册的学生。选课系统类模型如图 12-1 所示,其中 Person 是抽象类。

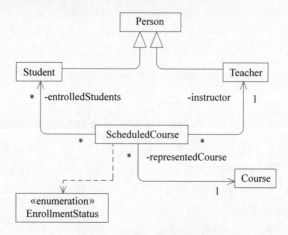

图 12-1 选课系统类模型

左下角是一个枚举类型 EnrollmentStatus,有四个枚举量注册成功(SUCCESS)、注册失败:已满员(SECTION_FULL)、注册失败:前驱课未修(PREREQ)、注册失败:已经注册(PREV_ENROLL)。

12.2 架构设计

采用数据访问对象(Data Access Object,DAO)模式作为基于文件持久化设施的选课系统架构。DAO 设计模式是一种软件设计模式,广泛用于以 Java 为平台的应用系统开发中。

目的是把具体的数据访问与业务功能分离。通过DAO接口让客户端代码仅仅依赖于抽象的接口，而不直接暴露具体操作持久化设施API或SQL语句。DAO设计模式的主要目标和优点如下。

（1）封装数据访问。DAO类封装了对磁盘数据的具体访问，这样可以隐藏底层实现细节，降低模块间的耦合度。

（2）易于切换不同的实现。如果未来需要更换数据库系统或者调整数据存储方式，只需在客户端代码中更换DAO的实现类即可，而无须更改使用DAO的业务逻辑代码。

（3）可复用。多个业务组件可以共享同一个DAO。

（4）便于单元测试。由于DAO提供了清晰的接口，因此可以在不依赖实际数据的情况下进行单元测试，只需要模拟DAO的行为即可。

图12-2展示了DAO架构。基于文件的控制台应用客户端通过DAO接口和实现类创建学生、教师、课程和课等对象，让这些对象交互完成选课的业务功能。将来如果使用某个数据库管理系统，如MySQL，只需更换实现类即可。

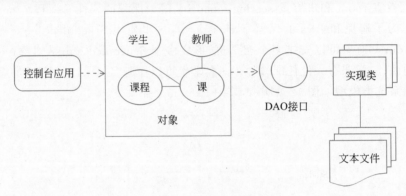

图12-2　DAO架构

基于文件的控制台应用通过DAO接口从磁盘文件读取学生、教师等各种实体的记录，创建为对象，并使用线性表或Map等集体对象管理。由于对Course和ScheduledCourse对象的查找操作频繁，所以使用Map对象进行管理；而Teacher和Student对象则放在线性表（List）中。线性表List中的元素有序、可重复；Map是"键-值"集合，其中的元素无序、Key值不能重复，value值可以重复。

12.3　详细设计

首先设计客户端代码如源文件12-1所示。这个客户端程序输出：

```
=========学生=========
张三 (201901001) [学士 - 数学]
李四 (201901002) [学士 - 计算机科学与技术]
王五 (201901003) [学士 - 计算机科学与技术]
=========教师=========
董永 (副教授,计算机科学与技术)
赵云 (教授,计算机科学与技术)
```

郭天（教授，数学）
=====课程 Course=====
CS103：数据结构与算法，3.0 学分
MAT101：概率与统计，3.0 学分
CS101：C 程序设计，3.0 学分
CS102：面向对象程序设计，3.0 学分
CS201：离散数学，3.0 学分
===开出的课 ScheduledCourse===
CS102-2，周四，下午 4:00-6:00，郭天（教授，数学），25
CS103-1，周一，下午 6:00-8:00，董永（副教授，计算机科学与技术），20
MAT101-1，周五，下午 4:00-6:00，赵云（教授，计算机科学与技术），15
CS101-1，周一，上午 8:00-10:00，郭天（教授，数学），30
CS101-2，周二，上午 8:00-10:00，赵云（教授，计算机科学与技术），30
CS102-1，周三，下午 2:00-4:00，董永（副教授，计算机科学与技术），25
CS201-1，周一，下午 4:00-6:00，郭天（教授，数学），1

学生 张三 试图注册 CS201-1
注册成功
学生 王五 试图注册 CS201-1
注册失败：已满员

源文件 12-1　AppFile.java

```java
1 package com.abc.file;
2
3 import com.abc.domain.Course;
4 import com.abc.domain.ScheduledCourse;
5 import com.abc.domain.Student;
6 import com.abc.domain.Teacher;
7 import com.abc.domain.EnrollmentStatus;
8
9 import com.abc.dao.CoursesDAO;
10 import com.abc.dao.TeachersDAO;
11 import com.abc.dao.ScheduledCoursesDAO;
12 import com.abc.dao.StudentsDAO;
13
14 import com.abc.dao.impl.file.CoursesDaoImpl;
15 import com.abc.dao.impl.file.TeachersDaoImpl;
16 import com.abc.dao.impl.file.ScheduledCoursesDaoImpl;
17 import com.abc.dao.impl.file.StudentsDaoImpl;
18
19 /**
20  * SRS File version
21  * @Author Dong D.
22  * @LastModified 2024-2-1
23  */
24 public class AppFile {
25
26     public static void main(String[] args) {
27         StudentsDAO students = new StudentsDaoImpl();
28         System.out.println("=========学生=========");
29         for (Student s : students.getAll()) {
30             System.out.println(s);
31         }
```

```java
        TeachersDAO teachers = new TeachersDaoImpl();
        System.out.println("=========教师=========");
        for (Teacher p : teachers.getAll()) {
            System.out.println(p);
        }

        CoursesDAO courses = new CoursesDaoImpl();
        System.out.println("=====课程 Course=====");
        for (Course c : courses.getAll()) {
            System.out.println(c);
        }

        ScheduledCoursesDAO scheduledCourses
            = new ScheduledCoursesDaoImpl(courses);


        //安排任课教师,根据工号查找教师对象,安排所任课
        scheduledCourses.getByID("CS101-1")
            .setInstructor(teachers.getByPID("123403"));
        scheduledCourses.getByID("CS101-2")
            .setInstructor(teachers.getByPID("123402"));
        scheduledCourses.getByID("CS102-1")
            .setInstructor(teachers.getByPID("123401"));
        scheduledCourses.getByID("CS102-2")
            .setInstructor(teachers.getByPID("123403"));
        scheduledCourses.getByID("CS103-1")
            .setInstructor(teachers.getByPID("123401"));
        scheduledCourses.getByID("MAT101-1")
            .setInstructor(teachers.getByPID("123402"));
        scheduledCourses.getByID("CS201-1")
            .setInstructor(teachers.getByPID("123403"));

        System.out.println("===开出的课 ScheduledCourse===");
        for (ScheduledCourse s : scheduledCourses.getAll()) {
            System.out.println(s);
        }
        System.out.println();

        //演示学生张三注册了一门容量为1的课,王五再注册则失败
        EnrollmentStatus status;

        System.out.println("学生 张三 试图注册 CS201-1");
        status = scheduledCourses.getByID("CS201-1")
            .enroll(students.getByName("张三"));
        System.out.println(status.value());

        System.out.println("学生 王五 试图注册 CS201-1");
        status = scheduledCourses.getByID("CS201-1")
            .enroll(students.getByName("王五"));
        System.out.println(status.value());


    }
}
```

第 27 行使用接口 StudentsDAO 声明了变量 students 来引用实现类 StudentsDaoImpl 对象；第 29 行调用实例方法 students.getAll() 返回所有的学生对象，然后逐一输出学生对象的字符串表示。

接下来按照类模型定义实体类 Student。然后为每个 Student 定义相应的数据访问接口 StudentDAO 及其实现类。把实体类 Student 声明在源文件 12-2 中。该类成员有全参数构造方法、getter 和 setter 方法，还有覆盖方法 toString、hashCode 和 equals。在源文件 12-1 第 30 行之所以能够输出 Student 对象的字符串表示，就是因为 println 方法调用了这里的 toString 方法。

源文件 12-2　Student.java

```java
 1 package com.abc.domain;
 2
 3 /**
 4  *
 5  * @author Dong
 6  * 2024-1-2
 7  * Student.java
 8  */
 9
10 public class Student extends Person {
11
12     private String major;
13     private String degree;
14
15     public Student(String name, String sid, String major, String degree) {
16         super(name, sid);
17
18         this.setMajor(major);
19         this.setDegree(degree);
20     }
21
22     public void setMajor(String major) {
23         this.major = major;
24     }
25
26     public String getMajor() {
27         return major;
28     }
29
30     public void setDegree(String degree) {
31         this.degree = degree;
32     }
33
34     public String getDegree() {
35         return degree;
36     }
37
38     /*
39      * 返回如下格式的字符串表示：<br> 张三 (201901001) [硕士 - 计算机科学与技术]
40      *
41      * @see hebtu.dd.Person#toString()
```

```
42      */
43     public String toString() {
44         return this.getName() + " (" + this.getID() + ") ["
45             + this.getDegree()   + " - " + this.getMajor() + "]";
46     }
47
48     @Override
49     public int hashCode() {
50         return this.getID().hashCode();
51     }
52
53     @Override
54     public boolean equals(Object obj) {
55         if (obj == null)
56             return false;
57         Student s = (Student) obj;
58         if (this.getID().equals(s.getID())) {
59             return true;
60         }
61         return false;
62     }
63 }
```

Student 类继承了 Person 类。把 Person 类声明在源文件 12-3 中。

源文件 12-3　Person.java

```
1 package com.abc.domain;
2
3 /**
4  * Person.java
5  * @author Dong
6  * 2023-03-2
7  *
8  */
9
10 public abstract class Person {
11
12     private String name;
13     private String pID;
14
15     public Person(String name, String ssn) {
16         this.setName(name);
17         this.setID(ssn);
18     }
19
20     public Person() {
21         this.setName("");
22         this.setID("");
23     }
24
25     public void setName(String name) {
26         this.name = name;
27     }
```

```
28
29      public String getName() {
30          return name;
31      }
32
33      public void setID(String pID) {
34          this.pID = pID;
35      }
36
37      public String getID() {
38          return pID;
39      }
40
41      @Override
42      public String toString() {
43          return "Person [姓名: " + name + ",证号: " + pID + "]";
44      }
45
46      public int hashCode() {
47          final int prime = 31;
48          int result = 1;
49          result = prime * result + ((pID == null) ? 0 : pID.hashCode());
50          return result;
51      }
52
53      public boolean equals(Object obj) {
54          if (this == obj)
55              return true;
56          if (obj == null)
57              return false;
58          if (getClass() != obj.getClass())
59              return false;
60          Person other = (Person) obj;
61          if (pID == null) {
62              if (other.pID != null)
63                  return false;
64          } else if (! pID.equals(other.pID))
65              return false;
66          return true;
67      }
68
69 }
```

然后在源文件 12-4 中创建接口 StudentsDAO。源文件 12-1 第 29 行调用的 getAll 方法声明在接口中。另外还声明了根据姓名返回学生对象的 getByName 方法。

源文件 12-4　StudentsDAO.java

```
1 package com.abc.dao;
2
3 import java.util.List;
4 import com.abc.domain.Student;
5
```

```
 6 /**
 7  *
 8  * @Author Dong D.
 9  * @LastModified 2023-3-2
10  * StudentsDAO.java
11  */
12 public interface StudentsDAO {
13
14     /*
15      * 返回所有学生对象
16      */
17     List<Student> getAll();
18
19     /*
20      * 根据姓名返回学生对象
21      */
22     Student getByName(String name);
23 }
```

接下来在源文件 12-5 中声明 StudentsDao 接口的实现类 StudentsDaoImpl。把三个学生的信息存储在一个文本文件 students.dat 中，字段之间使用制表符隔开，一个学生对应一条记录：

张三	201901001	数学	学士
李四	201901002	计算机科学与技术	学士
王五	201901003	计算机科学与技术	学士

第 22 行的构造方法从文本文件中逐行读取记录并创建为学生对象，并添加到线性表 students 中。由于所有的对象都在堆区，通过 public 的实例方法就可以访问这个线性表。第 39 行和第 44 行实现了接口中的 getAll 方法和 getByName 方法。第 49 行调用的 equals 方法就是 Student 类中的覆盖方法 equals。

源文件 12-5　StudentsDaoImpl.java

```
 1 package com.abc.dao.impl.file;
 2
 3 import java.util.ArrayList;
 4 import java.util.List;
 5 import java.util.ListIterator;
 6 import java.util.Scanner;
 7 import java.io.File;
 8 import java.io.IOException;
 9
10 import com.abc.dao.StudentsDAO;
11 import com.abc.domain.Student;
12
13 /**
14  *
15  * @author Dong D.
16  * @LastModified 2024-2-1
17  * StudentsDaoImpl.java
18  */
19 public class StudentsDaoImpl implements StudentsDAO {
```

```java
20      private List<Student> students = new ArrayList<>();
21
22      public StudentsDaoImpl() {
23          Student s1, s2, s3;
24          try (Scanner sc = new Scanner(
25              new File("D://javaWork//plainFiles//students.dat"))) {
26              while (sc.hasNext()) {
27                  String name = sc.next();
28                  String id = sc.next();
29                  String major = sc.next();
30                  String degree = sc.next();
31                  students.add(new Student(name, id, major, degree));
32              }
33          } catch (IOException e) {
34              throw new RuntimeException(e.getMessage());
35          }
36      }
37
38      @Override
39      public List<Student> getAll() {
40          return students;
41      }
42
43      @Override
44      public Student getByName(String name) {
45          Student s = null;
46          ListIterator<Student> iterator = students.listIterator();
47          while (iterator.hasNext()) {
48              s = iterator.next();
49              if (s.getName().equals(name))
50                  break;
51          }
52          return s;
53      }
54  }
```

类似地创建实体类 Tearcher、接口 TeachersDAO 及其实现类；创建实体类 Course，接口 CoursesDAO 及其实现类。实体类 ScheduledCourse 声明在源文件 12-6 中。第 20 行声明了引用授课教师的变量 instructor，显然这种设计约束了一门课至多一位授课教师。第 21 行使用 Map 对象管理选课名单，实现了 ScheduledCourse 与 Student 的多对多关联。除了构造方法、setter 和 getter 方法、toString 方法、hashCode 方法、equals 方法外，还声明了注册课程 enroll 和退课方法 drop，见第 119 行和第 136 行。第 145 行的方法 getTotalEnrollment 返回当前的选课人数；第 149 行的 getEnrolledStudents 方法返回选课名单。方法 isEnrolledIn 用来判断某个学生注册了这门课。

源文件 12-6 ScheduledCourse.java

```java
1   package com.abc.domain;
2
3   import java.util.HashMap;
4   import java.util.Map;
5
```

```java
 6  /**
 7   *
 8   * @author Dong D.
 9   * 2024-1-2
10   * ScheduledCourse.java
11   */
12  public class ScheduledCourse {
13
14      private int classNo;                                    //课号
15      private String dayOfWeek;                               //周几
16      private String timeOfDay;                               //时间
17      private String room;                                    //教室
18      private int seatingCapacity;                            //课容量
19      private Course representedCourse;                       //课程
20      private Teacher instructor;                             //任课教师
21      private Map<String, Student> enrolledStudents;          //选课名单
22
23      public ScheduledCourse(int sNo, String day, String time,
24              Course course, String room, int capacity) {
25          setClassNo(sNo);
26          setDayOfWeek(day);
27          setTimeOfDay(time);
28          setRepresentedCourse(course);
29          setRoom(room);
30          setSeatingCapacity(capacity);
31
32          setInstructor(null);
33          enrolledStudents = new HashMap<String, Student>();
34      }
35
36      public void setClassNo(int no) {
37          classNo = no;
38      }
39
40      public int getClassNo() {
41          return classNo;
42      }
43
44      public void setDayOfWeek(String day) {
45          dayOfWeek = day;
46      }
47
48      public String getDayOfWeek() {
49          return dayOfWeek;
50      }
51
52      public void setTimeOfDay(String time) {
53          timeOfDay = time;
54      }
55
56      public String getTimeOfDay() {
57          return timeOfDay;
58      }
```

```java
59
60      public void setInstructor(Teacher prof) {
61          instructor = prof;
62      }
63
64      public Teacher getInstructor() {
65          return instructor;
66      }
67
68      public void setRepresentedCourse(Course c) {
69          representedCourse = c;
70      }
71
72      public Course getRepresentedCourse() {
73          return representedCourse;
74      }
75
76      public void setRoom(String r) {
77          room = r;
78      }
79
80      public String getRoom() {
81          return room;
82      }
83
84      public void setSeatingCapacity(int c) {
85          seatingCapacity = c;
86      }
87
88      public int getSeatingCapacity() {
89          return seatingCapacity;
90      }
91
92      /**
93       * 返回如下格式：<br> CS101-1,周一,上午 8:00-10:00,赵教授,2
94       */
95      public String toString() {
96          return getRepresentedCourse().getCourseNo()
97                  + "-" + getClassNo()
98                  + ", " + getDayOfWeek()
99                  + ", " + getTimeOfDay()
100                 + ", " + getInstructor()
101                 + ", " + getSeatingCapacity();
102     }
103
104     /**
105      * 课程号(course no.)和班号(class no.)使用联合称为"完全号"。
106      * 例: "CS101-1"。
107      */
108     public String getFullScheduledClassNo() {
109         return getRepresentedCourse().getCourseNo()
110                 + "-" + getClassNo();
111     }
```

```java
112
113 /**
114  * 如果学生已经注册,返回 PREV_ENROLL,
115  * 如果容量满,则返回 SECTION_FULL,
116  * 把该学生加入到选课名单中,返回 SUCCESS。
117  * @ see EnrollmentStatus
118  */
119 public EnrollmentStatus enroll(Student s) {
120     EnrollmentStatus status = EnrollmentStatus.SUCCESS;
121     if (enrolledStudents.containsValue(s)) {
122         status = EnrollmentStatus.PREV_ENROLL;
123     } else {
124         if (enrolledStudents.size() < getSeatingCapacity()) {
125             enrolledStudents.put(s.getID(), s);
126         } else {
127             status = EnrollmentStatus.SECTION_FULL;
128         }
129     }
130     return status;
131 }
132
133     /**
134      * 退课
135      */
136     public boolean drop(Student s) {
137         if (! isEnrolledIn(s))
138             return false;
139         else {
140             enrolledStudents.remove(s.getID());
141             return true;
142         }
143     }
144
145     public int getTotalEnrollment() {
146         return enrolledStudents.size();
147     }
148
149     public Map<String, Student> getEnrolledStudents() {
150         return enrolledStudents;
151     }
152
153     public boolean isScheduledClassOf(Course c) {
154         if (c == representedCourse)
155             return true;
156         else
157             return false;
158     }
159
160     /**
161      * 判断是否已经选过本课
162      *
163      */
164     public boolean isEnrolledIn(Student s) {
```

```
165        if (enrolledStudents.values().contains(s))
166            return true;
167        else
168            return false;
169    }
170
171 }
```

类似地，新建文本文件 ScheduledCourses.dat，存储开出的课，放在 D:\javaWork\plainFiles 文件夹中：

```
1 CS101    1    周一    上午 8:00-10:00    A101    30
2 CS101    2    周二    上午 8:00-10:00    A202    30
3 CS102    1    周三    下午 2:00-4:00     C105    25
4 CS102    2    周四    下午 4:00-6:00     D330    25
5 CS103    1    周一    下午 6:00-8:00     E101    20
6 MAT101   1    周五    下午 4:00-6:00     D241    15
7 CS201    1    周一    下午 4:00-6:00     A205    1
```

然后设计接口 ScheduledCoursesDAO 及其实现类 scheduledCourses.dat。程序从这些文件中读取数据，创建对象，并使用线性表等集体管理这些对象。把所有用来创建对象所需的数据访问接口创建在包 com.abc.dao 中，各个接口的实现类 CoursesDaoImpl、TeachersDaoImpl、StudentsDaoImpl、ScheduledCoursesDaoImpl 在包 com.abc.dao.impl.file 中。

这些实现类分别创建并维护集体对象。例如 StudentsDaoImpl 类中维护存储若干学生对象的集体对象。都是通过读取文本文件中的记录创建对象，然后使用线性表或者 Map 等容器管理这些对象。实现类使用文件实例化了三位教师，三名学生和五门课程，并把这些对象分别由 List 对象和 Map 对象进行管理。然后根据时间、地点和课程容量创建某学期的七门课，并为这七门课安排了任课教师。通过 DAO 架构隔离了业务应用对外部数据的依赖，当改用数据库而不使用文件时就可以不改动现有的程序实现扩展。项目中源文件及其所在的文件夹如图 12-3 所示。

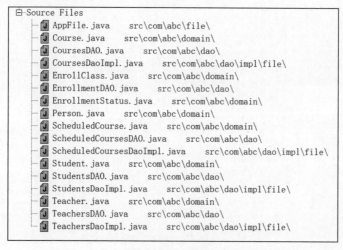

图 12-3　源文件

源文件 12-1 中的客户端代码第 50 行到第 63 行为每门课调用实例方法 setInstructor 安排授课教师。第 75 行演示向 CS201-1 课发送消息 enroll，参数是名字为张三的学生对象，即学生张三试图注册 CS201-1。由于当前的选课人数为 0，班容量是 1，所以注册成功；接下来学生李四继续注册该课，由于选课人数等于班容量，注册失败。

基于给定的源代码，可进行如下扩展：

（1）实现"等待队列"功能。即当因某门课因座位满而无法选上时将该学生放入等待队列，一旦有空位，则从队列中移出一位学生进行注册。程序应输出：

=========退课等待队列自动注册成功=======
学生 张三 试图退课 CS201-1
张三退课成功！
李四注册成功！
当前等待队列的学生为：[王五 (202001003) [学士 - 计算机科学与技术]]
当前选课名单为：{202001002=李四 (202001002) [学士 - 计算机科学与技术]}

（2）限制学生不能选重复的 ScheduledCourse 对象，也不能选重复的 Course 对象。程序应输出：

=========重复课程=========
学生 张三 试图注册 CS101-1
注册成功
学生 张三 试图注册 CS101-2
注册失败：重课
学生 张三 试图注册 CS101-1

附录 A Unicode Basic Latin 字符

表 A-1 列出了 Unicode Basic Latin 字符。

表 A-1 Unicode Basic Latin 字符

代码点 （十六进制）	代码点 （十进制）	英 文 名 字	中文名字	字符
0x0000	0	NULL	空	
0x0001	1	START OF HEADING	标题开始	
0x0002	2	START OF TEXT	正文开始	
0x0003	3	END OF TEXT	正文结束	
0x0004	4	END OF TRANSMISSION	传输结束	
0x0005	5	ENQUIRY	请求	
0x0006	6	ACKNOWLEDGE	确认	
0x0007	7	BELL	振铃	
0x0008	8	BACKSPACE	退格	
0x0009	9	HORIZONTAL TABULATION	水平制表	
0x000A	10	LINE FEED	换行	
0x000B	11	VERTICAL TABULATION	垂直制表	
0x000C	12	FORM FEED	换页	
0x000D	13	CARRIAGE RETURN	回车	
0x000E	14	SHIFT OUT	移位输出	
0x000F	15	SHIFT IN	移位输入	
0x0010	16	DATA LINK ESCAPE	数据链路转义	
0x0011	17	DEVICE CONTROL ONE	设备控制 1	
0x0012	18	DEVICE CONTROL TWO	设备控制 2	
0x0013	19	DEVICE CONTROL THREE	设备控制 3	
0x0014	20	DEVICE CONTROL FOUR	设备控制 4	
0x0015	21	NEGATIVE ACKNOWLEDGE	否认	
0x0016	22	SYNCHRONOUS IDLE	同步空闲	
0x0017	23	END OF TRANSMISSION BLOCK	传输块结束	
0x0018	24	CANCEL	取消	
0x0019	25	END OF MEDIUM	介质结束	
0x001A	26	SUBSTITUTE	替换	
0x001B	27	ESCAPE	转义	
0x001C	28	FILE SEPARATOR	文件分隔符	

续表

代码点 （十六进制）	代码点 （十进制）	英 文 名 字	中文名字	字符
0x001D	29	GROUP SEPARATOR	组分隔符	
0x001E	30	RECORD SEPARATOR	记录分隔符	
0x001F	31	UNIT SEPARATOR	单元分隔符	
0x0020	32	SPACE	空格	
0x0021	33	EXCLAMATIONPOINT	叹号	!
0x0022	34	DOUBLE QUOTATION MARK	双引号	"
0x0023	35	NUMBER SIGN	数字符	#
0x0024	36	DOLLAR SIGN	美元符	$
0x0025	37	PERCENT SIGN	百分号	%
0x0026	38	AMPERSAND	与	&
0x0027	39	SINGLE QUOTATION MARK/APOSTROPHE	单引号/撇号	'
0x0028	40	LEFT PARENTHESIS	左括号	(
0x0029	41	RIGHT PARENTHESIS	右括号	)
0x002A	42	ASTERISK/STAR	星号	*
0x002B	43	PLUS SIGN		+
0x002C	44	COMMA	逗号	,
0x002D	45	HYPHEN/MINUS	连字符	-
0x002E	46	PERIOD/DOT/FULL STOP	句号	.
0x002F	47	SLASH/SOLIDUS	斜线号	/
0x0030	48	DIGIT ZERO		0
0x0031	49	DIGIT ONE		1
0x0032	50	DIGIT TWO		2
0x0033	51	DIGIT THREE		3
0x0034	52	DIGIT FOUR		4
0x0035	53	DIGIT FIVE		5
0x0036	54	DIGIT SIX		6
0x0037	55	DIGIT SEVEN		7
0x0038	56	DIGIT EIGHT		8
0x0039	57	DIGIT NINE		9
0x003A	58	COLON	冒号	:
0x003B	59	SEMICOLON	分号	;
0x003C	60	LESS-THAN SIGN	小于号	<
0x003D	61	EQUALS SIGN	等于号	=
0x003E	62	GREATER-THAN SIGN	大于号	>

附录 A   Unicode Basic Latin字符

续表

代码点 （十六进制）	代码点 （十进制）	英 文 名 字	中文名字	字符
0x003F	63	QUESTION MARK	问号	?
0x0040	64	COMMERCIAL AT/AT SIGN	AT号	@
0x0041	65	LATIN CAPITAL LETTER A		A
0x0042	66	LATIN CAPITAL LETTER B		B
0x0043	67	LATIN CAPITAL LETTER C		C
0x0044	68	LATIN CAPITAL LETTER D		D
0x0045	69	LATIN CAPITAL LETTER E		E
0x0046	70	LATIN CAPITAL LETTER F		F
0x0047	71	LATIN CAPITAL LETTER G		G
0x0048	72	LATIN CAPITAL LETTER H		H
0x0049	73	LATIN CAPITAL LETTER I		I
0x004A	74	LATIN CAPITAL LETTER J		J
0x004B	75	LATIN CAPITAL LETTER K		K
0x004C	76	LATIN CAPITAL LETTER L		L
0x004D	77	LATIN CAPITAL LETTER M		M
0x004E	78	LATIN CAPITAL LETTER N		N
0x004F	79	LATIN CAPITAL LETTER O		O
0x0050	80	LATIN CAPITAL LETTER P		P
0x0051	81	LATIN CAPITAL LETTER Q		Q
0x0052	82	LATIN CAPITAL LETTER R		R
0x0053	83	LATIN CAPITAL LETTER S		S
0x0054	84	LATIN CAPITAL LETTER T		T
0x0055	85	LATIN CAPITAL LETTER U		U
0x0056	86	LATIN CAPITAL LETTER V		V
0x0057	87	LATIN CAPITAL LETTER W		W
0x0058	88	LATIN CAPITAL LETTER X		X
0x0059	89	LATIN CAPITAL LETTER Y		Y
0x005A	90	LATIN CAPITAL LETTER Z		Z
0x005B	91	LEFT SQUARE BRACKET	左方括号	[
0x005C	92	BACKSLASH	反斜线	\
0x005D	93	RIGHT SQUARE BRACKET	右方括号	]
0x005E	94	CIRCUMFLEX ACCENT	抑扬音符号	^
0x005F	95	UNDERSCORE/LOW LINE	下画线	_
0x0060	96	GRAVE ACCENT	沉音符	`

续表

代码点 （十六进制）	代码点 （十进制）	英 文 名 字	中文名字	字符
0x0061	97	LATIN SMALL LETTER A		a
0x0062	98	LATIN SMALL LETTER B		b
0x0063	99	LATIN SMALL LETTER C		c
0x0064	100	LATIN SMALL LETTER D		d
0x0065	101	LATIN SMALL LETTER E		e
0x0066	102	LATIN SMALL LETTER F		f
0x0067	103	LATIN SMALL LETTER G		g
0x0068	104	LATIN SMALL LETTER H		h
0x0069	105	LATIN SMALL LETTER I		i
0x006A	106	LATIN SMALL LETTER J		j
0x006B	107	LATIN SMALL LETTER K		k
0x006C	108	LATIN SMALL LETTER L		l
0x006D	109	LATIN SMALL LETTER M		m
0x006E	110	LATIN SMALL LETTER N		n
0x006F	111	LATIN SMALL LETTER O		o
0x0070	112	LATIN SMALL LETTER P		p
0x0071	113	LATIN SMALL LETTER Q		q
0x0072	114	LATIN SMALL LETTER R		r
0x0073	115	LATIN SMALL LETTER S		s
0x0074	116	LATIN SMALL LETTER T		t
0x0075	117	LATIN SMALL LETTER U		u
0x0076	118	LATIN SMALL LETTER V		v
0x0077	119	LATIN SMALL LETTER W		w
0x0078	120	LATIN SMALL LETTER X		x
0x0079	121	LATIN SMALL LETTER Y		y
0x007A	122	LATIN SMALL LETTER Z		z
0x007B	123	LEFT CURLY BRACKET	左花括号	{
0x007C	124	VERTICAL LINE/VERTICAL BAR	竖线	|
0x007D	125	RIGHT CURLY BRACKET	右花括号	}
0x007E	126	TILDE	波浪号	~
0x007F	127	\<control\>: DELETE	删除	

附录 B jGRASP 调试和展演

jGRASP(https://www.jgrasp.org/)是一个轻量级的集成开发环境,由奥本大学(Auburn University)开发,目的是用于程序设计语言教学。jGRAPS 不仅提供了基本的编辑、编译程序的功能,还集成了调试(debug)、展演(run in canvas)、源代码风格检查(checkstyle)、单元测试(jUNIT)等功能。本节介绍调试和展演的用法。

假设有计算器类 Calculator 提供了加减乘功能,并使用加法来实现乘法:

```
1  public class Calculator {
2      private int a;
3      private int b;
4
5      void setA(int x) {
6          a = x;
7      }
8
9      void setB(int y) {
10         b = y;
11     }
12
13     int add() {
14         return a + b;
15     }
16
17     int minus() {
18         return a + b;
19     }
20
21     int multiply(){
22         int i = 1;
23         int sum = 0;
24         while (i < b) {
25             sum = sum + a;
26             i = i + 1;
27         }
28         return sum;
29     }
30 }
```

类 Test 创建计算器对象并调用减法功能:

```
1  public class Test{
2      public static void main(String[] args){
3          Calculator c = new Calculator();
4          c.setA(10);
5          c.setB(3);
6          int result = c.minus();
7          System.out.println(result);
8      }
9  }
```

把光标放置在 main() 方法中，运行程序，发现程序的输出是 13，而不是期望的 7，如图 B-1 所示。

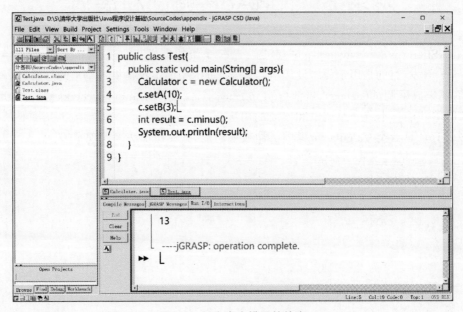

图 B-1　程序产生错误的输出

可以从第 6 行开始逐条语句执行程序，每执行一条语句都暂停程序执行，观察变量的取值是否正确来识别出错位置。下面按如下步骤寻找哪一行代码引发了错误。首先设置断点以让程序暂停(suspend)执行。

在源代码编辑窗口中，在打算设置断点的行(本例中是第 6 行)上把鼠标指针移动到左侧边框处，当指针形状变成红色八边形时，单击。红色八边形固定在该行左侧，标识此处有断点。再次单击红色八边形，取消断点，称为 toggle breakpoint，如图 B-2 所示。

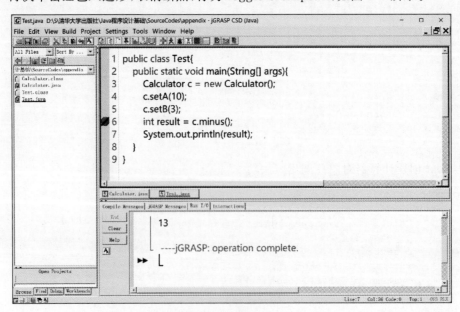

图 B-2　设置断点

附录 B jGRASP调试和展演

在工具按钮栏单击蜘蛛形状的"调试"工具按钮，进入调试运行状态，如图 B-3 所示。在调试状态下，左侧调试标签页中会出现调试按钮：。从左到右依次是单步（Step）、步进（Step in）、步出（Step out）、运行到光标处（Run to the cursor）、暂停运行（Pause）和继续运行（Resume）。

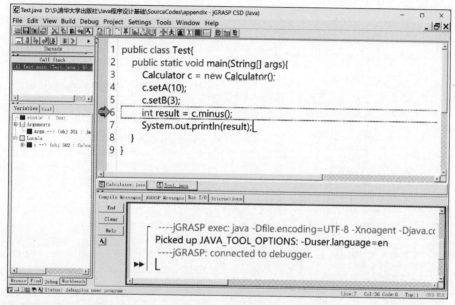

图 B-3 调试状态

如果单击调试控制按钮组最左侧的 Step 调试按钮则执行当前行，暂停在下一行。

单击调试控制按钮组左数第 2 个 Step in 调试按钮，如果当前行是方法调用，则进入方法体；否则同 Step 调试按钮。

如果单击调试控制按钮组左数第 3 个 Step out 调试按钮，则执行光标所处方法体中剩余行，返回调用处。

如果单击 Resume 调试按钮，则让程序继续运行，直到遇到下个断点。其他调试按钮依次为：Turn on/off auto step mode，Turn on/off auto resume mode（>>），Toggle on/off BYTE CODE SIZE STEP（⇓），Suspend new threads（仅用于调试多线程程序）。

因为第 6 行是一个方法调用 minus()，所以单击 Step in 调试按钮，进入方法 minus() 的方法体，如图 B-4 所示。

此时在 Variable 窗口观察变量的值。自上而下分别是当前对象（this）的成员变量（单击 this 左侧的加号"+"可展开）、参数和局部变量。审核第 18 行发现运算符应该是"-"而不是"+"。这就是程序运行结果错误的原因。找到原因后单击 Run I/O 窗口左侧的 End 按钮结束当前调试，回到编辑窗口修改源代码，重新编译 Calculator.java，重新把光标放置在main()方法中运行程序。

在 Variable 标签页右击变量，选择 View by Name 或者 View by Value 可打开 Viewer 窗口观察变量的内容。

在 Viewer 窗口右击某变量，选择 Watch for Access 紧盯某变量被访问的情况：设置 Watch 后 Resume（继续），当变量一旦被读写，则会暂停，观察变量值，然后再 Resume。

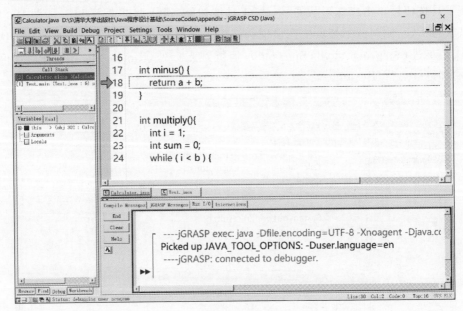

图 B-4 步进

假设调用方法 multiply()，程序的输出是 20 而不是期望的 30，如图 B-5 所示。

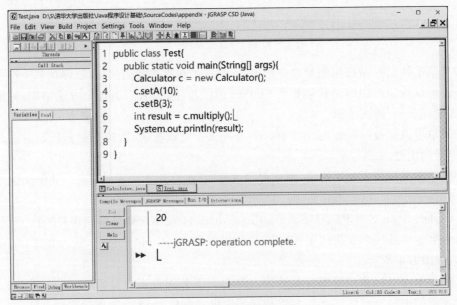

图 B-5 调用乘法功能

为了定位错误位置，首先在方法调用处设置断点，如图 B-6 所示。然后单击工具栏中的 Debug 按钮进入调试状态，再单击 Step in 调试按钮进入方法体。

单击 Step 按钮执行一条语句，然后暂停在下一条语句，如图 B-7 所示。单击 Variables 标签页中的当前对象 this 左侧的＋展开成员变量。连续单击 Step 按钮直到到达第 26 行，此时可观察到局部变量 i 的值是 1，sum 的值是 10。但是，由于处在循环体中，如果一直单击 Step 按钮不可行，可以让程序每循环一次在第 25 行暂停，观察了局部变量取值后再让程

图 B-6　设置断点

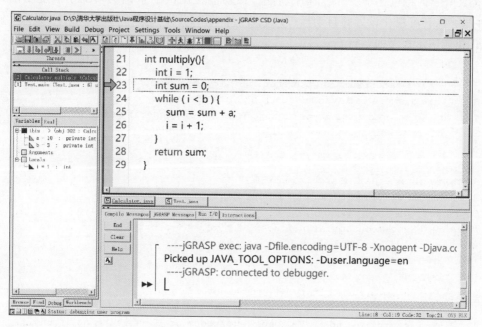

图 B-7　单步

序继续运行。首先在第 25 行设置断点,然后单击 Resume 调试按钮,程序会继续执行直到遇到下个断点。如图 B-8 所示。然后单击 Step 调试按钮,同时观察局部变量 i 和 sum 的变化,发现 i 取值 3,与成员变量 b 的值相等了,不满足循环条件,而此时 sum 的值是 20。也就是说,少循环一次。发现出现错误的位置后,单击 Run I/O 标签页中 End 按钮结束调试状态,回到编辑器修改程序,把循环条件 i<b 改为 i<=b。重新编译 Calculator,重新运行 main() 方法。

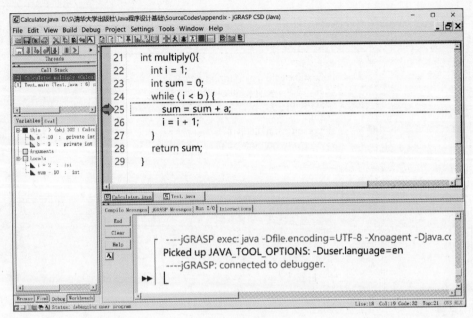

图 B-8 继续

jGRASP 的特色在于"展演"程序功能。该功能在调试的基础上增加了对变量的可视化,并能够保存可视化设置并自动运行。假设有如下二分查找算法的演示程序,那么可使用"展演"功能演示程序的运行。

```
1  /**
2   * BinarySearchExample with static binarySearch method.
3   * This example is intended to illustrate the viewer canvas
4   * with the Presentation Structure Identifier rendering of
5   * an array which has been "configured" with indexes.
6   */
7
8  public class BinarySearchExample {
9      public static void main(String[] args) {
10         int[] a = {12, 34, 56, 65, 73, 81, 97};
11         System.out.println("Index of 97 is: " + binarySearch(97, a));
12     }
13     /**
14      * binary search.
15      * @param key the binary search target.
16      * @param intArray the sorted array to be searched.
17      * @return the index of <code>key</code> or -1 if it was not found.
18      */
19
20     public static int binarySearch(int key, int[] array) {
21         int low = 0;
22         int high = array.length - 1;
23         int index = -1;
24         while (low <= high) {
25             int mid = low + (high - low) / 2;
```

```
26              if (key < array[mid]) {
27                  high = mid - 1;
28              } else if (key > array[mid]) {
29                  low = mid + 1;
30              } else {
31                  index = mid;
32                  break;
33              }
34          }
35          return index;
36      }
37  }
```

首先在桌面应用窗口的工具栏中单击 Run in Canvas 按钮 进入"展演"模式,如图 B-9 所示。"展演"模式在"调试"模式基础上增加了展板(Canvas)窗口。

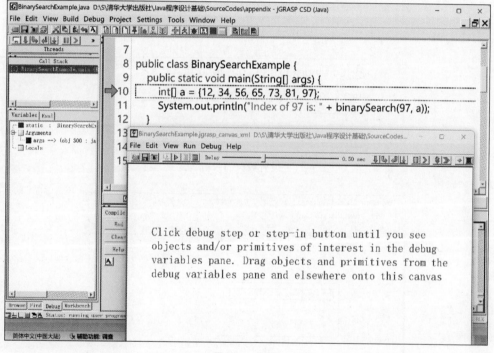

图 B-9　展演

单击"展板"窗口工具栏中的 Step 调试按钮,到第 11 行暂停;然后单击 Step in 调试按钮,进入方法调用 binarySearch() 的方法体。

连续单击 Step 调试按钮,使程序暂停在第 26 行,使得局部变量 low、high 和 mid 出现在 Variables 标签页中。然后把 low、high 和 mid 从 Variables 标签页拖放到"展板"窗口,如图 B-10 所示。然后单击数组对象,单击数组对象右上角的下列按钮,从下拉菜单中选择 Edit Index Expressions 菜单项,在 Edit Index Expressions for array 对话框中输入索引表达式"low♯mid♯high",表示把变量 low,mid 和 high 作为数组的索引变量,其中♯是分隔符,如图 B-11 所示。单击对话框中的 OK 按钮完成设置索引表达式,结果如图 B-12 所示。

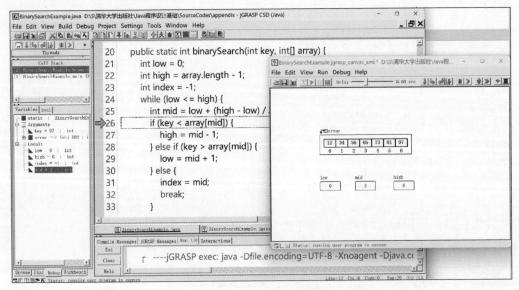

图 B-10 设置展示的变量

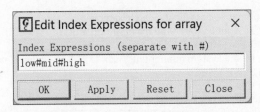

图 B-11 设置数组的索引表达式

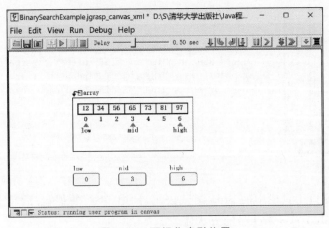

图 B-12 可视化索引位置

继续单击 Step 调试按钮,观察三个索引位置的变化,直到 low 和 high 相同,如图 B-13 所示。

单击 Step out 按钮结束 binarySearch()方法,返回到调用点;再次单击 Step out 按钮结束 main()方法。单击"展板"窗口中的 Save canvas 按钮,使用默认文件名保存展板文件。默认的扩展名是.jgrasp_canvas_xml。单击 Run I/O 标签页中的 End 按钮,停止"展演"

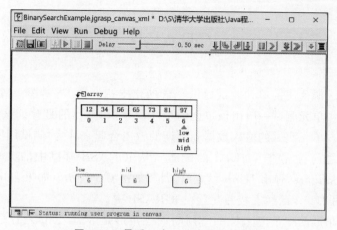

图 B-13　最后一次二分查找后的状态

模式。

　　需要演示的时候可以在"展板"窗口打开展板文件,然后单击 Run in viewer canvas 按钮,就可以单击 Step 调试按钮单步运行;或者单击 Turn autostep on and step in 按钮进入自动展演模式。此时可通过设置 delay 滑杆调整展演的速度。最后单击 End 按钮结束展演并关闭"展板"窗口。

附录 C　jGRASP 单元测试

JUnit 为 Java 源代码的单元测试提供了自动化支持。在 Java 中,一个方法、一个类都可视为一个单元。单元测试一般由该单元的程序设计执行,目的是验证该单元对于输入产生了预期的输出。单元测试的输入数据和预期的输出合起来称为"测试用例"。

使用 JUnit 之前首先查看 JUnit 的配置。从 jGRASP 窗口中的菜单 Tools 中选择 JUnit,再选择 Configure,弹出 JUnit 设置对话框,如图 C-1 所示。如果 JUnit Home 下拉列表框不为空,表明 JUnit 已经安装并配置在 jGRASP 中。

例如给定三角形的三条边 a、b 和 c,判断该三角形是等边三角形、等腰三角形、还是任意三角形? 这三种三角形类别的关系如图 C-2 所示(维恩图)。等边三角形也是等腰三角形、等腰三角形也是任意三角形。纬恩图中的三个不相交区域构成了三个有效等价类。最外边的大椭圆外面的区域表示非三角形。如果给定的三条边中任意一条小于 0,或者存在一条边的边长大于其余两条边边长之和,那么 a、b、c 无法构成三角形。这两种情况属于无效等价类。为每个无效等价类设计一个测试用例;为每个类别标签(也就是各种可能的输出)设计一个测试用例。

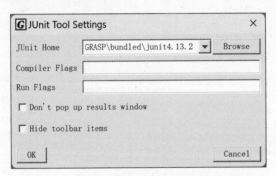

图 C-1　JUnit 设置对话框

图 C-2　三角形的分类

源文件 C-1　Triangle.java

```
1
2  /** Determines if three doubles can be sides of triangle. */
3  public class Triangle  {
4
5      /** Length of side 1. */
6      private double a;
7      /** Length of side 2. */
8      private double b;
9      /** Length of side 3. */
10      private double c;
11
12      /**
13       * Creates a Triangle object.
14       * @param a length of a side.
15       * @param b length of another side.
```

```
16        * @param c length of the third side.
17        */
18       public Triangle(double a, double b, double c) {
19           a = a;
20           b = b;
21           c = c;
22           if (a <= 0 || b <= 0 || c <= 0) {
23               throw new IllegalArgumentException("Sides: " + a + ", " + b
24                   + ", " + c
25                   + ": each one must be greater than zero.");
26           }
27           if ((a >= b + c) || (b >= a + c) || (c >= a + b)) {
28               throw new IllegalArgumentException("Sides: "
29                   + a + ", " + b + ", " + c
30                   + ": NOT a triangle.");
31           }
32       }
33
34       /**
35        * Classifies a triangle based on the lengths of the three sides.
36        * @return the triangle classification label.
37        */
38       public String classify() {
39           String result;
40           if ((a == b) && (b == c)) {
41               result = "equilateral";
42           } else if ((a != b) && (a != c) && (b != c)) {
43               result = "scalene";
44           } else {
45               result = "isosceles";
46           }
47           return result;
48       }
49   }
```

类Triangle定义完成后保存在源文件Triangle.java中。然后从菜单栏中选择Tools|JUnit|Creat Test File，jGRASP编辑窗口自动打开JUnit生成的测试驱动类TriangleTest，如源文件C-2所示。

源文件C-2　TriangleTest.java

```
1 import org.junit.Assert;
2 import org.junit.Before;
3 import org.junit.Test;
4
5
6 public class TriangleTest {
7
8       /** Fixture initialization (common initialization for all tests). **/
9       @Before public void setUp() {
10       }
11
12
13       /** A test that always fails. **/
```

```
14    @Test public void defaultTest() {
15        Assert.assertEquals("Default test added by jGRASP. Delete "
16            + "this test once you have added your own.", 0, 1);
17    }
18 }
```

其中,注解@Before是所有注解@Test公用的初始语句。把@Test注解的方法改为测试无效等价类的测试,测试用例[(−2,5,10),抛出异常],如源文件C-3所示。expected=IllegalArgumentException.class表示在运行这个测试用例时期望抛出IllegalArgumentException异常。然后从菜单栏中选择Tools|JUnit|Compile and Run Tests,运行测试,在Run I/O窗口会看到消息:Completed 1 tests 1 passed,表示完成了1个JUnit测试,通过了1个,即100%通过。jGRASP还会弹出图C-3所示的窗口,可视化地报告测试通过情况:左侧是完成情况,右侧是通过情况。

源文件 C-3　TriangleTest.java

```
1 import org.junit.Assert;
2 import org.junit.Before;
3 import org.junit.Test;
4
5
6 public class TriangleTest {
7
8     /** Fixture initialization (common initialization for all tests). **/
9     @Before public void setUp() {
10    }
11
12    /** Checks for bad arguments. */
13    @Test(expected = IllegalArgumentException.class)
14    public void argumentTest1() {
15        Triangle t = new Triangle(-2, 5, 10);
16    }
17 }
```

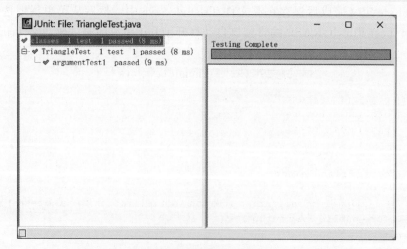

图 C-3　JUnit 结果窗口:完成的测试和通过的测试

继续补充对测试用例[(2,−5,10),抛出异常],[(2,5,−10),抛出异常]的测试,如源文件 C-4 所示。

源文件 C-4　TriangleTest.java

```java
1 import org.junit.Assert;
2 import org.junit.Before;
3 import org.junit.Test;
4
5
6 public class TriangleTest {
7
8     /** common initialization for all tests. **/
9     @Before public void setUp() {
10    }
11
12    /** Checks for bad arguments. */
13    @Test(expected = IllegalArgumentException.class)
14    public void argumentTest1() {
15        Triangle t = new Triangle(-2, 5, 10);
16    }
17    @Test(expected = IllegalArgumentException.class)
18    public void argumentTest2() {
19        Triangle t = new Triangle(2, -5, 10);
20    }
21    @Test(expected = IllegalArgumentException.class)
22    public void argumentTest3() {
23        Triangle t = new Triangle(2, 5, -10);
24    }
25 }
```

测试的输出是：Completed 3 tests 3 passed，表示测试了 3 个，通过 3 个。

添加测试用例［(2,5,10)，抛出异常］：

```java
/** Checks for "Not a triangle". */
@Test (expected = RuntimeException.class)
public void notTriangleTest() {
    Triangle t = new Triangle(2, 5, 10);
}
```

为每个分类标签添加一个测试用例：

```java
/** Checks for "Equilateral" triangle. */
@Test
public void equilateralTest1() {
    Triangle t = new Triangle(12, 12, 12);
    Assert.assertEquals("\nSides: " + 12 + " " + 12 + " " + 12,
        "equilateral", t.classify());
}

/** Checks for "Isosceles" triangle. */
@Test public void isoscelesTest1() {
    Triangle t = new Triangle(12, 12, 13);
    String result = t.classify();
    Assert.assertEquals("\nSides: 12, 12, 13", "isosceles",  result);
}
```

```
/** Checks for "Scalene" triangle. */
@Test public void scaleneTest1() {
    Triangle t = new Triangle(1, 2, Math.sqrt(2));
    Assert.assertEquals("\nSides: " + 1 + " " + 2 + " " + Math.sqrt(2),
        "scalene", t.classify());
}
```

运行测试,结果如图 C-4 所示,有 2 个测试未通过。然后从菜单栏中选择 Tools|JUnit|Compile and Debug Tests,选择进入调试模式,在未通过测试的方法前设置断点,单步运行查找错误位置。会发现源文件 C-1 Triangle.java 的第 19 行应为 this.a=a;第 20 行和第 21 行也应补充 this 关键字,重新编译 Triangle.java,再把光标移动到 TriangleTest.java 编辑窗口中重新运行测试。

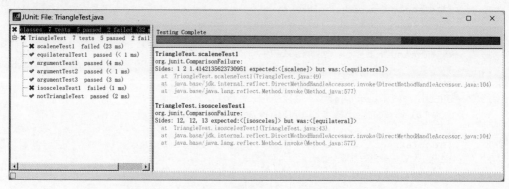

图 C-4　JUnit 结果窗口:部分测试未通过

Assert.assertEquals()用来断言期望的输出与实际的输出应该相等。如果不相等则抛出异常。该异常被 jGRASP 捕获后在 Run I/O 输出文本汇总信息以及在 JUnit 结果窗口中显示异常的栈跟踪信息。Assert.assertEquals()的第 1 个参数是抛出异常时显示错误提示消息。

当断言整数类型的期望值时的用法也类似:

```
Assert.assertEquals(errorMessage, expected, actual);
```

当断言浮点数类型的期望值时,还需要指定参数 delta 以避免舍入误差。该方法按照 Math.abs(expected-actual)<=delta 进行比较。例如:

```
assertEquals(aDoubleValue, anotherDoubleValue, 0.001)
```

当断言数组对象时使用:

```
assertArrayEquals(expected, actual)
```

assertNull("Theactual result should be null", actual)断言空;assertNotNull 断言不空。

assertTrue 断言真;assertFalse 断言假。例如:

```
assertTrue("5 is greater than 4", 5 > 4)
```

详细的 API 查阅 https://junit.org/junit4/javadoc/latest/org/junit/Assert.html。

当一个方法或者一个类中的所有方法完成单元测试后,才能提交,供其他模块使用。

参 考 文 献

[1] Oracle. The Java Tutorials[EB/OL]. 2024-1-29. https://docs.oracle.com/javase/tutorial/.
[2] Java Examples. Java Code Examples[EB/OL]. 2023-12-2. https://www.javacodeexamples.com/.
[3] Refsnes Data. Java Tutorial[EB/OL]. 2023-12-7. https://www.w3schools.com/java/default.asp.
[4] 阿里云开发者社区. 阿里巴巴 Java 开发手册(终极版)[EB/OL]. 2024-1-28. https://developer.aliyun.com/ebook/386/92064.
[5] 董东. 文本分析途径的课程持续改进目标识别[J]. 软件导刊,2023,22(2):132-135.
[6] 黑马程序员. Java 基础入门[M]. 第 3 版. 北京:清华大学出版社,2022.
[7] 耿祥义,张跃平. Java 2 实用教程(题库+微课视频版)[M]. 第 6 版. 北京:清华大学出版社,2021.